FORTRAN FOR ENGINEERING PHYSICS
Mechanics, Data Analysis, and Heat

FORTRAN FOR
ENGINEERING PHYSICS
Mechanics, Data Analysis, and Heat

ALAN B. GROSSBERG

Associate Professor of Physics
Chairman, Engineering Science Division
The University of Wisconsin-Parkside

McGRAW-HILL BOOK COMPANY

New York   St. Louis   San Francisco   Düsseldorf   Johannesburg   Kuala Lumpur   London
Mexico   Montreal   New Delhi   Panama   Rio de Janeiro   Singapore   Sydney   Toronto

FORTRAN FOR ENGINEERING PHYSICS
Mechanics, Data Analysis, and Heat

Library of Congress Catalog Card Number 77-160707

07-024971-7

1 2 3 4 5 6 7 8 9 0 W H W H 7 9 8 7 6 5 4 3 2 1

*This book was set in Baskerville and News Gothic by York Graphic Services, Inc., and
printed and bound by The Whitlock Press, Inc.*
*The designer was Elliot Epstein; the drawings were done by John
Cordes, J. & R. Technical Services, Inc. The editors were Bradford
Bayne and Barry Benjamin. John A. Sabella supervised production.*

CONTENTS

PREFACE

There is hardly any academic field to be found today which has not been profoundly influenced by the development of the digital computer. The computer is used extensively in all fields of physical sciences and engineering, as well as in business, the social sciences, and statistics. Even such seemingly unlikely areas as English literature and music have not remained untouched by the computer revolution.

It is with these ideas in mind that this textbook has been written. It represents a fundamental departure from the usual general physics laboratory manual. The book introduces digital computing techniques at the outset and makes use of them in conjunction with experiments similar to the traditional general physics experiments. Subjects receiving heavy emphasis throughout the book, in addition to important physical principles and experimental procedures, are computer processing of experimental data and numerical computing methods. Each numerical method is introduced in an appropriate place in the text where it provides some definite advantage for the treatment of the experimental data.

The laboratory equipment required for the performance of these experiments is the conventional complement of general physics laboratory apparatus. In most cases, no modification of this equipment is required. Only the method of performing the experiments and the way in which the experimental data is handled depart from the traditional.

This book has been written in two volumes. Each volume is self-contained and may be used independently. The experiments described in the book follow closely the usual sequence of subjects covered in a standard two-semester general physics course. The first volume deals with mechanics and heat, as well as elementary data-analysis techniques, including the treatment of experimental errors. The second volume, in preparation, will cover optics, electricity, magnetism, and modern physics, and will also treat various data-analysis procedures.

The mathematical background required for these experiments begins at the precalculus level; the mathematical level increases as the book progresses. Experiments 1 through 9 of the first volume require no knowledge of calculus for their understanding, or for the computer analyses which are associated with them. Thus the first part of the text lends itself readily to a course with which calculus is taken concurrently, or even to a course which has no calculus requirement at all.

The last four experiments in the first volume are somewhat more demanding mathematically. They involve the concepts of elementary differential equations, least-squares data analysis, and numerical integration, all of which are covered in the text. By the time these experiments are performed, the student should have encountered the ideas of derivative, integral, and the elements of differential equations in his calculus course. Any of these last four experiments may be omitted, however, without any loss of continuity.

No background in computer programming is presumed for students who use this textbook. Details of FORTRAN programming are presented step by step; they are discussed in a separate programming section in each experiment. The early experiments in the book, in particular, are designed to illustrate the fundamental principles of FORTRAN programming. Afterward, the emphasis is placed on various numerical computational methods.

Extensive use of flow charts is made in this text. The idea of a flow chart is first presented in Experiment 2. Thereafter, in the programming section of each experiment, there appears a flow chart of a program to perform the analysis of the experimental data. It is the job of the student to write an appropriate FORTRAN program, "debug" it, and then analyze his data with its help. A printout of each final error-free program and of all computed results ordinarily should accompany every completed laboratory report.

Although this book deals exclusively with FORTRAN (both FORTRAN II and FORTRAN IV), all of the calculations discussed here may be performed equally well using other programming languages and with most of the small general-purpose laboratory computers now widely available. For example, all of the computations described in Experiments 1–13 have been programmed in FOCAL, a FORTRAN-like interactive language developed by Digital Equipment Corporation, and carried out on their small general-purpose PDP-8 computer. The flow charts of the corresponding programs are identical to the ones presented here.

The second volume begins with experiments that review basic FORTRAN programming concepts. It then proceeds to discuss such topics as numerical solution of transcendental equations, least-squares data analysis, computer solution of potential problems, solution of systems of simultaneous equations, and more material on differential equations. Among the additional programming procedures covered in this part of the text are subprograms, computer treatment of complex numbers, and computer graphing methods.

I should like to acknowledge many helpful discussions with the late Professor R. A. Jaggard and Professor J. M. Martin of the University of Wisconsin—Milwaukee. I am grateful to Professor Hugh D. Young of Carnegie-Mellon University for his careful review of the manuscript of this book and for the many excellent suggestions he made during the course of its preparation. Acknowledgment should be made to the University of Wisconsin Computing Center for the use of computing facilities while these experiments were being developed.

I must mention with gratitude the ideas derived for this book from numerous students in general physics and computer programming courses over the last several years. Finally, I would like to express my appreciation to my wife for her assistance with the proofreading of the manuscript, and for her patience and encouragement while the writing was in progress.

Alan B. Grossberg

EXPERIMENT 1
FUNDAMENTALS OF
FORTRAN PROGRAMMING

1. INTRODUCTION

Fortran is a widely used computer language originally developed at IBM for the formulation and solution of mathematical and scientific problems. It has been continuously improved and extended since its introduction and is now applied to the solution of many problems in business and the social sciences as well. Fortran is a universal computer language in the sense that it is not dependent on the particular characteristics of any one computer, and with very few exceptions a Fortran program written for one machine will work equally well with any other. Furthermore, Fortran is oriented toward the scientist and the types of problems that he ordinarily encounters and not the detailed machine operations performed by the computer in effecting the solution. In fact, the user need not be at all familiar with the operation and construction of a digital computer in order to be an effective Fortran programmer.

By a Fortran *program* is meant a sequence of instructions (*statements*) written in the Fortran language, which is a mixture of English words, numbers, mathematical symbols, and several special keyboard characters, such as * and /. This set of instructions is loaded into the computer along with the data that is to be processed by the program before any actual computations are performed. Prior to the computation process, the computer must itself *translate* the statements of the Fortran program into a sequence of machine language instructions. These machine instructions are much simpler and many times more numerous than the statements contained in the program. The translation is accomplished with the aid of another machine language program that has previously been loaded into or is permanently stored in the *memory* of the computer. The latter program is referred to as a Fortran *processor* or *compiler*. The translation process is termed *compilation*.

The Fortran program is usually loaded into the computer by means of punched cards or tape which are read by a card or tape reader. A deck of cards containing the Fortran program state-

1

ments, punched one to a card, is called the *source* deck. The translated program, which appears as a sequence of machine-language instructions, is known as the *object* program. It may be stored in the memory of the computer, or it may be punched out on a deck of cards called the *object* deck (or it may be written out on a tape).

The actual computation begins after the object program has been loaded into the computer (if it was not stored in memory) along with the data upon which the program operates. The performance of these machine instructions is known as the *execution* of the program. In most large-scale computer installations the extra steps involved in the production of an outputted object program deck are omitted, and the execution of the program may begin immediately after the compilation has been completed.

In this first experiment, we are going to examine some of the elementary features of Fortran programming as they apply to a simple computational problem. In order to provide a concrete example for this study, one which can be related to an easily performed laboratory experiment, we will compute the value of the gravitational acceleration constant g from data obtained with a simple pendulum. The period of such a pendulum is simply related to its length and g. Therefore, if its length and period are measured, g can be determined from the values which are obtained.

2. APPARATUS The simple pendulum consists of a small weight, which is acted upon by the force of gravity, attached to the lower end of a light rod or cord of length l. The small suspended weight is known as a *bob*. The upper end of the cord is attached to a fixed support. The simple pendulum diagram in Fig. 1-1 indicates the support,

FIG. 1-1
Simple pendulum

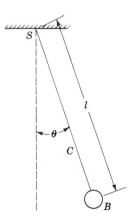

cord, and bob by the letters S, C, and B. The angle between cord and vertical is labeled θ. It must not exceed $10°$ if we are to be able to apply the standard formula for the period of a pendulum as a function of its length, Eq. [1-1] below. Otherwise, correction terms are required, and a simple analysis is ruled out.

The support S must be adjustable so that the length of cord C may be varied. It usually takes the form of a chuck or a clamp which may be loosened to allow the cord to be pulled through it and set to a desired length. The period of vibration of the pendulum is obtained by timing some regular number of vibrations (10 or 20, for example) and then dividing the measured time interval by that number.

In order to obtain a more accurate value of the period of vibration of the simple pendulum, a greater number of oscillations should be timed. This improves the accuracy of the determination in two ways. First, the time interval will be a number with greater significance. For example, if a stopwatch with a minimum interval of 0.01 sec is available, by timing a single oscillation a period of, say, 1.50 sec might be measured. Ten oscillations might require 15.03 sec. One hundred oscillations might give a measured interval of 150.32 sec, etc. The period determined in this manner will have one more significant digit each time the number of vibrations timed is multiplied by a factor of 10; i.e., the values obtained for the period of vibration in the above examples would be 1.50, 1.503, and 1.5032 sec. Each of these numbers contains one more significant digit than the preceding one.

The relative effect of the finite reaction time of the experimenter is also minimized as the number of vibrations timed is increased. Let us assume that an uncertainty of 0.05 sec due to the reaction time of the operator occurs each time the stopwatch is started or stopped. The total uncertainty in the measured time interval attributable to operator reaction time is then 0.1 sec. Expressed as a percentage in the three examples of the preceding paragraph, the reaction time uncertainty is, respectively, 6.7, 0.67, and 0.067 percent of the total interval timed. In terms of the period of vibration, these percentages correspond to uncertainties of 0.1, 0.01, and 0.001 sec. Under these circumstances, in order to achieve an experimental accuracy of 1 percent or better, at least 7 vibrations would have to be timed.

3. THEORY

When the amplitude of vibration of a simple pendulum is restricted to angles of $10°$ or less, its period may be expressed as a function of its length to within 0.2 percent by

[1-1] $T = 2\pi\sqrt{\dfrac{l}{g}}$

In Eq. [1-1], g is the acceleration due to gravity. This relation is proved in the general physics references of the bibliography. Equation [1-1] may be solved for g explicitly in terms of T and l

[1-2] $g = 4\pi^2\dfrac{l}{T^2}$

If higher accuracy is needed than 0.2 percent when the pendulum is vibrating with an amplitude in the neighborhood of $10°$, or if the amplitude of vibration is greater than $10°$, it is necessary to add correction terms to Eqs. [1-1] and [1-2]. It can be shown that with one correction term Eq. [1-1] becomes[1]

[1-3] $T = 2\pi\sqrt{\dfrac{l}{g}}\left(1 + \dfrac{1}{16}\theta_0^{\,2} + \cdots + \right)$

In Eq. [1-3], θ_0 is the maximum angular displacement of the pendulum. It must be expressed in *radians*. This formula is useful for angles θ_0 up to about $20°$. As an exercise, calculate the percentage increase in period caused by the correction term for θ_0 equal to $20°$.

When the correction term is included in the expression for T, the formula for g, Eq. [1-2], must also be modified to take account of it. This is most easily accomplished with the aid of the binomial formula

[1-4] $(1 + x)^2 = 1 + 2x + x^2$

This expression converges rapidly when the magnitude of x is much less than 1. If we set $x = (\theta_0/4)^2$ in Eq. [1-3], square the equation, apply the binomial formula, Eq. [1-4], and retain only the linear term in x, we obtain

[1-5] $T^2 \approx 4\pi^2\dfrac{l}{g}(1 + x)^2 = 4\pi^2\dfrac{l}{g}\left[1 + 2\left(\dfrac{\theta_0}{4}\right)^2\right]$

Solving Eq. [1-5] explicitly for g, we have

[1-6] $g = 4\pi^2\dfrac{l}{T^2}\left[1 + 2\left(\dfrac{\theta_0}{4}\right)^2\right]$

[1] C. Kittel, W. D. Knight, and M. A. Ruderman, "Berkeley Physics Course—Volume I—Mechanics," p. 227, McGraw-Hill Book Company, New York, 1965. See also F. W. Sears and M. W. Zemansky, "University Physics," 4th ed., p. 167. Addison-Wesley Publishing Company, Inc., Reading, Mass., 1970.

In a Fortran program discussed at length in the next section, either Eq. [1-2] or [1-6] may be used to compute g from the experimental period-versus-length data.

4. PROGRAMMING A Fortran program consists of an ordered sequence of instruction statements which are loaded into a computer, translated into machine language, and then executed in order.[1] In this first discussion of programming techniques, we will be primarily concerned with just two kinds of Fortran statements, *input-output* and *arithmetic* statements. In order to write a complete program, we will also require a FORMAT statement, which is one example of a *specification* statement. It will be briefly described here, the main discussion being deferred to a later experiment in which input-output programming procedures are considered in more detail.[2]

The simplest kind of computational Fortran program consists of the following program segments:

1   Input of data

2   Arithmetic steps

3   Output of results (and data, if desired)

4   Specification (Format) statements

These program segments are arranged in the order in which they will appear in our completed program. It should be pointed out that the position of specification statements such as FORMAT in a Fortran program is not critical. For convenience, we have chosen to place the FORMAT statement near the end of the program. The FORMAT is included in a program to supply the computer with information on how the input data will appear on punched data cards, and in what form to produce the output data in the printout. Other specification statements such as DIMENSION (Experiment 4) must be located at the beginning of a program.

[1] As will be discussed in future experiments, a very important exception occurs when, due to the decision-making feature of Fortran, the computer is made to return to an earlier statement or to jump forward in the program and skip some statements.

[2] The programming section of Experiment 7 will be concerned with input-output procedures and a detailed discussion of FORMAT.

The remainder of this section will be devoted to a discussion of the kinds of statements referred to in the outline above. Each will be treated in a separate subsection. Finally, the program of Fig. 1-3, which illustrates the use of these statements in a practical situation, will be discussed in detail. We begin the discussion with a consideration of Fortran expressions.

Fortran Expressions   A Fortran expression is basically an algebraic expression. It consists of mathematical quantities, such as $2x$ and π, which are manipulated by the standard arithmetic operations of addition, subtraction, multiplication, division, and exponentiation. The latter operation, the raising to a power of a number or expression, is treated in Fortran as a distinct operation. A simple example of an algebraic expression is $2\pi r(r + l)$. It will be recognized as the total outer surface area of a right circular cylinder. An expression such as this may be evaluated for many different pairs of the algebraic *variables* r and l, an infinite number of pairs, in fact. In general, a variable may assume many different values, often an infinite number. The definite numbers in the area expression, π and 2, are referred to as *constants*. Their values do not change.

Fortran expressions, like their algebraic counterparts, are composed of variables, constants, and operation symbols. A Fortran constant is a definite number whose value does not change during the execution of a program, while a Fortran variable may take on many different numerical values. The five basic Fortran arithmetic operations and the symbols that represent them are

Operation	Symbol
Addition	+
Subtraction	−
Multiplication	*
Division	/
Exponentiation	**

The name of a Fortran variable must begin with a capital English letter. It may otherwise consist of any combination of capital letters and numerical characters 0 to 9. Lowercase letters are not used in Fortran programming. The maximum length that may be assigned to the name of a Fortran variable is six characters. A Fortran variable may be represented by a single character. In this case, the character must be some capital letter.

Fortran expressions resemble algebraic expressions very closely. Parentheses are used in the same manner. Fortran variable names, numerical constants, and operation symbols are similar to corresponding algebraic quantities. For example, we could write a Fortran expression for the surface area of a right circular cylinder, $2\pi r(r + l)$, as

6.28319*R*(R + L)

The algebraic expression for the area might also be written $\pi r(2r + 2l)$. The corresponding Fortran expression would be

3.14159*R*(2.*R + 2.*L)

We notice in the last example several differences between algebraic expressions and their Fortran equivalents. First, multiplication in Fortran expressions must be indicated *explicitly* with an asterisk. No symbol need be inserted between quantities that are to be multiplied together in an algebraic expression.

Numerical constants are also different. In Fortran they must be written as rational numbers of finite length. Thus, π, an irrational number is rounded off in the above expression to five decimal places. Otherwise the algebraic expression and its Fortran equivalent are very similar.

Fortran notation is very flexible. It allows many different forms of the same expression. As a simple illustration of this flexibility, we may rewrite the preceding Fortran expression for the surface area of a cylinder using radius and length variables named RADIUS and LENGTH

3.14159*RADIUS*(2.*RADIUS + 2.*LENGTH)

There is one more very significant difference between the last Fortran expression and the corresponding algebraic one. The constant factor of *two,* which was written simply as 2 in the algebraic expression, shows up in the Fortran expression as 2., a number that contains a decimal point. Such a number is known as a *floating-point* constant. The number 2 written without a decimal point is a *fixed-point* (or *integer*) constant. Fixed- and floating-point numbers are handled in entirely different fashion in a digital computer, so it is very important to distinguish between them in all Fortran expressions. Fixed-point numbers may only take on positive or negative integer values (or zero). Floating-point numbers do not have any such restriction imposed on them. They need only be rational numbers in decimal form (such as 12.34, or -0.567×10^8).

In general, it is not permissible to mix fixed- and floating-point numbers in a single Fortran expression. An important exception is an integer exponent in a floating-point expression, which *is* allowed. In order to ensure that the numerical values which are assumed by variables in a floating-point expression be treated as floating-point numbers, these variables must be *defined* as floating-point variables. This is accomplished, as discussed below, either by convention, or through the use of a *type statement,* in this case, a REAL statement.

Fortran variables and Fortran constants are either of fixed- or floating-point form. Only floating-point variables may be used in floating-point expressions (with the exception of variables that appear in exponents, which might be of either type). There is an important convention in Fortran regarding the naming of variables. In the absence of type-specification statements (such as REAL and INTEGER, which will be discussed in a later experiment), the variable type is determined by the first letter in its name. Any variable beginning with I, J, K, L, M, or N is treated by Fortran as an integer, or fixed-point, variable. All others are handled as floating-point variables. Thus, RADIUS would be floating point, and LENGTH fixed point. In order that the expression for the area of a cylinder, as written above, be entirely floating point, the variable LENGTH would have to be defined as a floating-point variable.[1]

A word on the use of nested parentheses in arithmetic expressions is in order here. It is usual in algebraic expressions to indicate nested pairs of parentheses by a sequence of different symbols, such as parentheses, brackets, and braces. For instance, we might have

$$\left\{ \left[\left(\frac{f}{f_0} \right) - \left(\frac{f_0}{f} \right) \right]^2 + \left(\frac{1}{Q} \right)^2 \right\}^{1/2}$$

In Fortran, all expressions must be written without fractions, subscripts, and superscripts, as such, on a single typewritten line. Also, only one form of parenthesis symbol is available. The Fortran equivalent of the preceding algebraic expression would be

$$(((F/FO) - (FO/F))**2 + (1./Q)**2)**0.5$$

When nested parentheses appear in a Fortran expression, all indicated arithmetic operations are performed in sequential order, starting with the innermost pair of parentheses. At the far left of the expression above there are three consecutive left-hand parentheses. The innermost of these (i.e., the *rightmost* of the three) is completed by the nearest right-hand parenthesis, the one following F/FO. Thus, the division of F by FO is performed first.

---

[1] This could be accomplished with an appropriate REAL statement, as will be discussed in detail in Experiment 4.

The center parenthesis of the three is closed by the right-hand parenthesis immediately preceding the **2 operation. It causes the evaluation of F/FO − FO/F, *before* the squaring is performed.

The outermost (leftmost) of the three adjacent left-hand parentheses is associated with the right-hand parenthesis that follows the (1./Q)**2 term. It requires that the latter term *next* be added to the result obtained in the preceding paragraph. *After* this has been done, exponentiation to the 0.5 power is performed.

In the absence of parentheses, there is a definite order in which arithmetic operations are performed in Fortran: exponentiation, followed by multiplication and division, and, finally, addition and subtraction. This eliminates the need for many pairs of parentheses. In our example, making use of this rule, we could write the preceding Fortran expression as

((F/FO − FO/F)**2 + 1./Q**2)**0.5

It should be verified that there is no ambiguity in this expression regarding the order in which arithmetic operations are to be performed.

Arithmetic Statements

Before we can proceed with a discussion of arithmetic statements, we must give attention to the manner in which variables are handled in Fortran. Each variable must first be introduced or defined in some statement in a program. This statement causes a numerical value to be read in or computed for the variable. The numerical value is stored in a specific memory location which Fortran assigns to the variable. Thereafter, whenever the given variable appears in any arithmetic expression, Fortran causes its numerical value to be retrieved from that memory location and substituted for the variable in the arithmetic expression. The value of the variable itself is not changed by this retrieval process. In this fashion, each variable and each arithmetic expression is given a definite *numerical* value in Fortran. It is these numerical values which are operated upon in arithmetic statements.

Mathematical processes in Fortran are performed by means of arithmetic statements. The general form of an arithmetic statement is

Fortran variable = Fortran expression

The equals sign in such a statement has a very special and important meaning. It requires that all arithmetic operations included in the expression on the right-hand side of the statement first be performed, using the current values of all of the variables which appear in the expression. Fortran then causes the numerical value computed for the expression to be stored in the memory location assigned to the variable named on the left-hand side of the equals sign. The previous value of this variable is erased first (and lost). This special meaning attached to the equals sign in Fortran permits statements to be written that would be nonsensical if interpreted as algebraic equations. An example is

$$I = I + 1$$

This arithmetic statement, written in fixed point, causes the variable I to be increased by 1, and the result to be stored as I. It accomplishes the incrementing of the variable I, and is a frequently used arithmetic statement.

One more property of the Fortran arithmetic statement should be pointed out. The two sides of the arithmetic statement need not be of the same type; one may be integer and the other floating point. This feature of Fortran programming is often used for converting one kind of variable into another. A statement that will cast the integer value of the fixed-point variable N into floating-point form is

$$XN = N$$

The floating-point variable XN may then be used in floating-point expressions without further modification. If the above statement were written in reverse

$$N = XN$$

whatever value XN had at the time of execution (not necessarily integral) would be converted[1] to an integer and stored as N.

Input-output Statements
We will be concerned primarily with input and output operations performed by card readers and line printers. The common Fortran input and output statements which are used with these input-output devices are READ, WRITE, and PRINT. Associated with each such statement is a numbered FORMAT statement. Before we proceed with our discussion of input-output statements, however, we must consider the conventions Fortran imposes on punched statement cards, and how to prepare a card acceptable to a Fortran compiler.

[1] The conversion occurs by *truncation* (chopping off) of all digits to the right of the decimal point, leaving an integer.

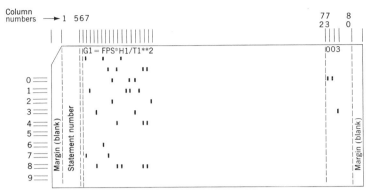

FIG. 1-2
Typical Fortran statement
card. The statement itself
is punched anywhere in
columns 7 through 72.
Columns 73 to 80 are
ignored by the Fortran
compiler and may be used
for card identification
numbers or characters.
Statement numbers (if
needed) are punched in
columns 1 to 5. Row
numbers are indicated at
the left of the card.

Figure 1-2 illustrates a Fortran statement card which has been provided with column numbers to indicate the proper punching of a single Fortran statement. Notice that the card contains 80 numbered columns. Each column consists of twelve rows, the lower ten rows being numbered 0, 1, 2, . . . , 9. When a letter, number, or other keyboard character is punched, one or more rectangular holes is cut in certain rows of *one* column. These punches are the coded representation of that character. When the card passes through a card reader, the punched holes are scanned electrically or optically, causing electrical impulses to be sent to the computer indicating the presence of the character. The numbers 0 through 9 have the simplest representation. Each consists of a single punch in the row associated with the number.

Statement numbers are needed for various reasons in a Fortran program to identify particular statements in the program. One example of a statement that must be numbered is the FORMAT statement. As will be discussed below, each of the basic input and output statements must be associated with a FORMAT statement to inform the computer how to treat all input-output data. The statement number is punched in columns 1 through 5, which are

11

specifically reserved for statement numbers.[1] It is punched with its least significant digit (e.g., the digit 3 in the number 123) in column 5. A two digit number would be punched in columns 4 and 5 etc. For example, a simple FORMAT statement numbered 100 might be punched on a statement card as follows:

```
| | |1|0|0| |F|O|R|M|A|T| |(|I|5|)| |
```

The number 100 appears in columns 3, 4, and 5. The meaning of the symbol I5 will be discussed later.

Ordinarily, only a small number of statements in a program need be assigned numbers. If a statement is unnumbered, columns 1 through 5 are left blank. Statement numbers in a program need not be consecutive, nor even in ascending or descending order. They are completely arbitrary, except that no two statements may be given the same number, and no statement may be assigned a number larger than 99999 (less than that in some of the smaller computer systems).

Column 6 of a Fortran statement card is usually left blank (except when a statement is being continued on the card, as discussed below). The statement itself is punched in columns 7 through 72. Blank spaces in a statement *between* words and operation symbols are ignored by the Fortran compiler. Any number of blanks may be used. All of the characters making up a given word (or number) must be written next to one another without any blank spaces between characters. Otherwise, the word will be interpreted as several distinct words. This will usually cause an error message to be printed by the computer, and will probably result in termination of the program.

It is not necessary to include blank spaces in a Fortran statement. In the interest of legibility, however, it is advisable to do so. As an example of the kind of compact notation that results when a statement is written without blanks, let us write a statement to compute the area A of a right circular cylinder as follows:

A=3.14159*R*(2.*R+2.*L)

---

[1] An exception to this rule is column 1. It may also be used to indicate to the compiler the presence on the card of a comment rather than a Fortran statement. This is accomplished by punching the letter C in column 1. The remainder of the card is then ignored by the Fortran compiler. Whatever message is punched on this card, however, will be printed out when the program statements of the source deck are listed by the printer. Comment cards are very useful for labeling various segments of a program for future reference. Examples of comments of this kind will be found in the program of Fig. 1-3.

This statement would be made more readable by the insertion of several blank spaces (e.g., before and after the equals sign).

Sometimes there is not enough space available in columns 7 through 72 of a punched card to accommodate a particular Fortran statement. In this case, the statement may be continued on a second card after punching the number 1 in column six of that card. If a third card is needed, the number 2 is punched in its column 6 and the statement is continued etc. In Fortran IV any nonzero character may be punched in column 6 to indicate continuation. Columns 73 through 80 are ignored by the Fortran compiler. These columns may be used to provide identification numbers or symbols for the card, if desired.

READ statements are very similar in the two common versions of Fortran, Fortran II, and Fortran IV. The major difference between them is the inclusion of a number that indicates the reading device being employed along with the FORMAT statement number in a Fortran IV READ statement. The number 5 is often used to designate a card reader. READ statements which accomplish the reading of values of three floating-point variables A, B, and C and one fixed-point variable I from a single data card are:

```
READ 100, A, B, C, I     Fortran II
READ (5,100) A, B, C, I     Fortran IV
```

Each of these READ statements refers to a FORMAT statement numbered 100. The FORMAT describes the manner in which the data are to be found on the data card.

A more detailed discussion of input-output and FORMAT statements will be presented in Experiment 7. In order that we may begin immediately to write simple Fortran programs, we will now briefly describe a basic type of FORMAT statement that may be used in this and the following several experiments. It has the general form:

```
FORMAT (Specifications)
```

In parentheses after the word FORMAT itself are one or more *specifications* or *format codes*. Each specification refers to a variable in the READ statement; i.e., the first specification refers to the variable A in either of the above READ statements, the second to B, etc. In the first several experiments (through Experiment 6), only F and I specifications will be written. Floating-point numbers will be read in and printed out under F specifications; fixed-point numbers, under I specifications.

A sample FORMAT statement which might be used in conjunction with either of the above READ statements is

100 FORMAT (F10.4,F10.4,F10.4,I5)

It could be written in more compact notation as

100 FORMAT (3F10.4,I5)

What this statement does is alert the computer to expect three floating-point numbers punched in adjacent 10-column-wide blocks (or *fields*) followed by a single fixed-point number punched in the next five columns. The first F10.4 field begins in column 1 of the data card, the second in column 11, etc. The I5 field begins in column 31 and ends in column 35. It is possible to separate these data fields with blank columns, if desired. This may be accomplished with appropriate X specifications, as will be discussed in the programming section of Experiment 7.

Each of the F10.4 specifications indicates a floating-point number with four digits to the right of the decimal point. The number is located at the right-hand end of its 10-column-wide field. The I5 specification refers to an integer number with its least significant digit (e.g., the digit 3 in the number 123) located in the rightmost column of its five-column-wide field. The numbers 0.1234, 5.678, 90., and 123, punched on a data card as required[1] by either of the above FORMAT statements, would appear as

| bbbbb.1234 | bbbb5.678b | bbb90.bbbb | bb123 | bbbbb... |

The letter b indicates the presence of a blank (unpunched) column on the data card. Vertical lines are used to separate the three 10-column F10.4 fields and the single I5 field. They do not actually appear on the punched data card, of course.

It is not necessary to punch a decimal point in a number that is to be read under an F specification. The same numbers would be inputted with either of the above FORMAT statements if the numbers were punched on a data card in the following way

| bbbbbb1234 | bbbbb5678b | bbbb9bbbbb | bb123 | bbbbb... |

---

[1] Considerable latitude is given to a keypunch operator by the following Fortran convention regarding data read under an F specification. *The actual position of the decimal point in a number read under an F specification takes precedence over the location of decimal digits specified by the F code.* Thus, the number 0.1234 when punched *anywhere* in its 10-column field and read under an F10.0 specification, would still be interpreted as 0.1234.

In both of these examples, blank spaces are treated as zeroes. Thus, the number bbbb9bbbbb, read under an F10.4 specification, is interpreted as 90.0000 (zeroes to the *left* of the most significant digit in a number are ignored). Blank spaces to the *right* of the last nonzero character (i.e., the least significant digit) in a fixed-point number in an I field are also inputted as zeroes. For instance, if the number bbb1b were read in under an I5 specification, it would be interpreted as the number 10.

Fortran output statements are very similar to input statements. The FORMAT conventions are essentially the same. In input, the data contained on a single data card are referred to as one *record*. An output record is one line of line printer output. The simple type of FORMAT statement discussed previously causes a single output record to be produced; i.e., it causes one line of output to be printed. One line may consist of as many as 130 or more characters.

The F specification has the same significance in output as it does in input. For instance, an F10.4 specification produces a floating-point number located at the right of a 10-column field with four digits to the right of the decimal point. The I specification likewise has the same meaning as it did for input. The number 10, for example, printed under an I5 specification would appear in its field as bbb10.

The common output statements in Fortran II and Fortran IV are

```
PRINT 100, A, B, C, I     Fortran II
WRITE (6,100) A, B, C, I     Fortran IV
```

In addition to the number of its associated FORMAT statement, the Fortran IV WRITE statement contains a number that indicates the particular output device which is required. The number 6 in this example refers to a line printer.

The same type of FORMAT statement may be used both on input and output. In fact, it is possible to refer both input and output statements to a single FORMAT. The FORMAT statement numbered 100 in our example of data input would be an acceptable FORMAT to use with either of the PRINT or WRITE statements above, also. Try to write the line of output that the line printer would produce under these circumstances.

A Fortran Program   A Fortran program which calculates *g* from five pairs of period-versus-length data taken from a simple pendulum experiment is given in Fig. 1-3. It begins by reading in values of data pairs, each of which is punched on a separate data card under a 2F15.5

FIG. 1-3
Fortran program to com-
pute the average value of
the gravitational constant
g from simple pendulum
data.

```
C      COMPUTATION OF ACCELERATION DUE TO GRAVITY
C
C   READ IN DATA
       READ (5,100) H1, T1, H2, T2, H3, T3, H4, T4, H5, T5
C   COMPUTE ACCELERATION OF GRAVITY
       FPS = 4.0*3.14159**2
       G1 = FPS*H1/T1**2
       G2 = FPS*H2/T2**2
       G3 = FPS*H3/T3**2
       G4 = FPS*H4/T4**2
       G5 = FPS*H5/T5**2
       GBAR = (G1 + G2 + G3 + G4 + G5)/5.0
C   PRINT OUT DATA AND RESULTS
       WRITE (6,100) H1, T1, H2, T2, H3, T3, H4, T4, H5, T5, G1, G2,
     1 G3, G4, G5, GBAR
100 FORMAT (2F15.3)
       END
```

format specification. These data pairs are assigned to the Fortran variables T1 and H1, T2 and H2, etc. The value of $4\pi^2$ is then computed in an arithmetic statement and assigned to the variable FPS (representing "four pi squared"). Next, a value of g is computed for each of the five data pairs. These values of g are assigned to the Fortran variables G1, G2, G3, G4, and G5. In the final arithmetic statement, an average of these values of g is taken and stored as GBAR.

The input data pairs and the computed results, including all values of g and their average, are printed out with the aid of a WRITE statement and the same FORMAT (statement 100) that was used with the input READ statement. The FORMAT statement is located *almost* at the end of the program. The reason for saying "almost" is that every Fortran program must include after all its instructions (statements) an END statement. This is a signal to the Fortran compiler that no more statements remain to be translated. Data cards are included *after* the END card along with any *control* cards which your computer facility requires.[1] A sample output produced by the program of Fig. 1-3 is given in Fig. 1-4.

FIG. 1-4
Output generated by the
Fortran program of Fig.
1-3. Only numerical output
is produced by this pro-
gram; the literal labels
had to be added after-
wards to the printout.

H1	.200	T1	.898
H2	.400	T2	1.269
H3	.600	T3	1.554
H4	.800	T4	1.795
H5	1.000	T5	2.007
G1	9.791	G2	9.806
G3	9.809	G4	9.802
G5	9.801	GBAR	9.802

---

[1] Details of required control cards will be furnished by your computer installation.

5. PROCEDURE   **a**   Set up the simple pendulum apparatus illustrated in Fig. 1-1. Adjust the length of the pendulum to 20 cm. Time at least 20 or 30 vibrations as accurately as possible. Record the period and length of the pendulum. Repeat the period determination for pendulum lengths l of 40, 60, 80 cm, etc. Keep θ below $10°$.

b   Punch out the program of Fig. 1-3 on 80-column cards as described in Section 4. Punch the period-versus-length data in pairs on data cards according to the 2F15.5 format specification, i.e., as floating-point numbers with decimal points in 15-column fields. Have your program checked out before having it run. If possible, have a printout made of your program deck and data cards. Check it for spelling and punctuation errors. Next, have the program compiled, and, if error-free, executed. Include in your experiment report both a printout of the program and of the data and results.

c   As a variation of this experiment, repeat part a for a vibration amplitude of $20°$. Write a Fortran program that makes use of Eq. [1-6] to compute the value of g for each data pair. Note that the value of θ_0 (in radians) will have to be read in, or else included in an arithmetic statement explicitly. Compare the results of this program with those found in part b.

EXPERIMENT 2
ERRORS OF MEASUREMENT

1. INTRODUCTION Whenever a physical quantity is measured in the laboratory, there is always some error accompanying the determination. An error of measurement may be defined as the amount by which a given measurement differs from the true value of the physical quantity being determined. The *true* value of any quantity may be defined in turn as the average of many measurements of that quantity by an ideal observer equipped with perfect instruments. There are a number of reasons why one individual measurement may be subject to error. These are discussed in the following paragraphs.

Errors of measurement may be divided generally into two broad classes, systematic and random errors. Systematic errors occur when the observer, the measuring instrument, or the technique of measurement is biased in some way. There is generally one underlying cause for each systematic error. For instance, an incorrectly calibrated voltmeter might yield potential values all 5 percent too high. This is an example of a systematic *instrumental* error. A systematic *personal* error might arise from an observer reading a meter at some angle other than vertical. This is illustrated in Fig. 2-1. The correct reading is obtained when the meter is read from a point directly above the pointer, P. The correct reading is R. However, due to the angle θ at which the pointer is viewed aginst the meter scale, a different reading, R', is ob-

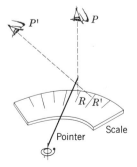

FIG. 2-1
Illustrating the effect of parallax in reading a meter

tained. This apparent shift of position of the meter pointer due to the angle at which it is viewed is known as *parallax*. Elimination of parallax from various measurements is a frequently occurring experimental problem.

When repeated determinations of a physical quantity are made with the same equipment, systematic errors will always show up in the same way. For example, a voltmeter might consistently read too high on the upper part of its scale and too low on the lower portion of the scale. Systematic errors tend to be of rather sizable magnitude and ordinarily can be correlated with one associated experimental factor. In any good laboratory determination, considerable care is taken to eliminate or minimize all systematic errors that might influence the value of the experimental result.

However, even after precautions have been made to reduce the effect of all major sources of personal, instrumental, and environmental errors, there still remain small variations among the values obtained in repeated trials of the experiment. These variations are due to a multitude of small errors of undetermined origin. Each of the small errors, if it could be traced, would be found to be due to some systematic disturbance of the experimental conditions, such as a temperature or pressure fluctuation etc. There are too many of them, and each is of too small a magnitude to be determined individually. Furthermore, each of these sources of error is just as likely to produce a positive variation of the experimental quantity being measured as a negative one. Such deviations from the true value of the physical quantity are known as *random* errors.

It should be pointed out that not all collections of random deviations from a mean value are symmetrical about the mean. An important exception occurs in nuclear physics. When low-level radioactive sources or cosmic ray particles are counted over fairly long time intervals, under suitable experimental conditions, small numbers of particles are counted in each interval. When the mean of a large number of counts is calculated, on a counts-per-interval basis, the number of counts per interval *below* the mean value by a given amount is different from the number of counts per interval *above* the mean by the same amount. This is true even when very good data are taken, and the discrepancy cannot be attributed to instrumental inaccuracy. Data of this kind follow a statistical law, formulated by Poisson, which is different from the law obeyed by the random errors of measurement that we are considering. In our experimental determinations, errors of measurement will be symmetrically distributed about their mean, if the number of measurements, N, is large. We will not pursue the subject of Poisson statistics further here.[1]

Intuitively, we are able to describe many of the characteristics possessed by a set of random measurement errors. If, as discussed above, we define the error of a particular measurement as the deviation of that measurement from the true value, which is the average of a large number of measurements, then a set of *random* errors would be expected to:

1   Cluster about the mean value, i.e., exhibit a most frequently occurring error value of zero

2   Exhibit a frequency of occurrence that decreases continuously as the error magnitude increases in either direction from zero

3   Have vanishingly few errors as the magnitude of error increases toward infinity, i.e., have no very large errors

4   Exhibit symmetry about zero error, i.e., show as many positive errors of a given magnitude as negative errors of that magnitude

The mathematical theory of random errors does, in fact, verify these predictions. The statistical function that describes how many measurements, out of a total number N, are expected to be found with each value of error is called an error *distribution*. When plotted, it is a *distribution curve*. The particular distribution curve predicted mathematically for a large collection of random error measurements is known as a *normal* or *gaussian* distribution curve. Figure 2-2 illustrates a typical gaussian distribution curve. It will be discussed further in Section 3. The reader should verify that this distribution has all the characteristics we previously intuitively suggested should be exhibited by a distribution of random errors.

We will be concerned in this experiment with the analysis of errors that occur during the course of physical measurements. Such errors accompany all experiments. We assume that precautions have been taken to eliminate all systematic errors and that only random errors remain in our experimental data. How to estimate the size of these random errors and how these errors affect the value calculated for an experimental result are topics that will be considered in Section 3. Our objective is to gain a knowledge of how to perform an analysis of the errors present in our experi-

[1] A good elementary discussion of Poisson statistics is given in chap. 5 of N. C. Barford, "Experimental Measurements: Precision, Error and Truth," Addison-Wesley Publishing Company, Inc., Reading, Mass., 1967.

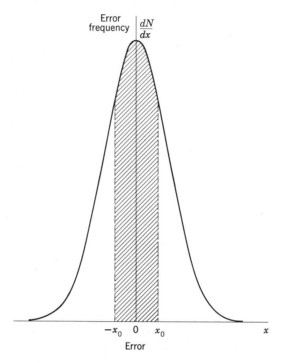

FIG. 2-2
Gaussian or normal error distribution curve. The theoretical equation for the Gaussian distribution is: $f(x) = (1/\sqrt{2\pi})\exp(-x^2/2)$, where x is the error in simplified relative-error units.

ment, and, on the basis of this error analysis, how much precision to attach to the calculated answer. In particular, we are concerned with the problem of how many figures should be included with the reported result and how to compute a value for the uncertainty that is to be reported along with it.

2. APPARATUS   In order to base the discussion of error-analysis techniques on a simple, concrete experiment, and in order to become familiar with several basic measuring instruments, we will determine directly the densities of several regularly shaped solid objects. Each object should be a homogeneous solid and one whose density may be found listed in the standard tables.[1] The density in each case will be determined with the aid of the definition

[2-1] $$\text{Density} = \frac{\text{mass}}{\text{volume}}$$

[1] Tables of the densities of various solids will be found in any edition of the "Handbook of Chemistry and Physics," Chemical Rubber Publishing Company, Cleveland, and the "American Institute of Physics Handbook," McGraw-Hill Book Company, Inc., New York, 1963, among others.

The mass of an object, which appears in the numerator of Eq. [2-1], may be determined either with a platform balance or with a chemical-type analytical balance. Details of the construction and operation of these instruments will be found in Appendix A. The basic difference between the two instruments is one of sensitivity. A good platform balance might permit mass determinations to the nearest 0.1 g, or, possibly, 0.01 g. An analytical balance should allow measurements to at least an accuracy of 0.1 mg. An estimate of the minimum sensitivity of the particular platform balance or analytical balance which is used in this experiment should be made before the experiment is actually performed. This may be done by noting the response of either balance when its lightest *rider* is moved a small distance, corresponding to a small increment of added mass. The smallest added mass which causes a definite, perceptible movement of the balance is a measure of its sensitivity.

The denominator of Eq. [2-1], the volume of the solid sample whose density is being determined, is calculated from various dimension measurements made with either a vernier caliper or micrometer caliper. The principal difference between the two instruments is again one of sensitivity. An ordinary vernier caliper permits length determinations to the nearest 0.1 mm, whereas with a micrometer caliper an accuracy of 0.01 or 0.001 mm might be attained. These instruments are also described in Appendix A.

3. THEORY   We begin our discussion of the results of the mathematical theory of random errors with a precise definition of what we mean by the term error. We are assuming that all sources of systematic error have been eliminated from our experiment. Under these circumstances, the true value of the physical quantity being measured is understood to be given by the average of a large number of repeated measurements of the quantity. Let each measurement be denoted by m_i and the mean value of all N measurements by $\bar{m}$. The error of the ith measurement is then defined by the following relation, in which the error is indicated by x_i:

[2-2] $x_i = m_i - \bar{m}$

The mean value, $\bar{m}$, is calculated by means of the standard formula

[2-3] $\bar{m} = \dfrac{1}{N} \displaystyle\sum_{i=1}^{N} m_i$

23

It is a good idea to digress for a moment here to explain the summation notation used in Eq. [2-3]. It will also be employed whenever needed in all succeeding experiments. The capital Greek letter *sigma*, Σ, is used to indicate summation. The summation index i runs from 1 to N, as indicated. The quantities being summed are the m_i. If Eq. [2-3] were to be written out explicitly, it would read

$$\bar{m} = \frac{1}{N}(m_1 + m_2 + \cdots + m_i + \cdots + m_N)$$

Sometimes, the index i, and its minimum and maximum values, are left out of the summation notation, and must be understood from the accompanying text, or from the meaning of the summation.

Strictly speaking, the quantity x_i defined in Eq. [2-2], which is called the *error* of the ith measurement, represents the *deviation* of the ith measurement (from the mean of the N measurements). However, since we are assuming that N is a large number, and that no systematic errors remain in our determinations, the mean of the N measurements, $\bar{m}$, and the true value should be the same. Therefore, we will use the terms error and deviation synonymously, although, in a rigorous sense, x_i should be termed the deviation of the ith measurement from the mean in all cases.

Error Distributions We must now examine the error distribution curve of Fig. 2-2 in greater detail. The curve is a plot of a *continuous* function. It describes the frequency with which each value of error occurs. To be able to interpret this kind of continuous distribution curve, we must first have a clear understanding of how a finite collection of error measurements can be represented by a bar-graph distribution "curve," or *histogram*. To illustrate how such a histogram of errors could be constructed, let us consider a hypothetical laboratory experiment.

Suppose that a large number of measurements of the length of a machined steel block have been made with a very precise micrometer. For the sake of simplicity, we will assume that the micrometer can be read to the nearest *micron* ($1\ \mu = 0.001$ mm) and that tenths of a micron can be estimated. This is much better accuracy than we would ordinarily expect to achieve, even with a very good precision micrometer.

We could compute the average of these measurements with Eq. [2-3], and then, using Eq. [2-2], the error associated with each measurement. Counting the number of measurements that correspond to an error of 0 μ (i.e., whose errors fall between -0.5 and $+0.5$ μ), we could draw a bar of that height, 1 μ wide, and centered at the origin of our graph. Counting the number of errors between 0.5 and 1.5 μ and constructing a bar of that height centered at 1.0 μ etc., we would obtain a bar-graph error distribution similar to Fig. 2-3.

When the number of measurements is large, the histogram that represents their distribution of errors should follow the normal error distribution curve quite closely. This close agreement between histogram and error curve would not be expected in the case of a small number of measurements, say 10. In our example, we are assuming that a very large number of measurements have been performed, and we would expect that the bar graph of Fig. 2-3 should conform closely to the gaussian error curve of Fig. 2-2.

A continuous distribution curve, like Fig. 2-2, may be considered to be a limiting case of a bar graph describing a finite collection of measurement errors. A continuous distribution results when the number of errors is increased to infinity and the width of the bars is shrunk to zero. It is really a plot of the number

FIG. 2-3
Histogram error distribution for a finite set of length measurements

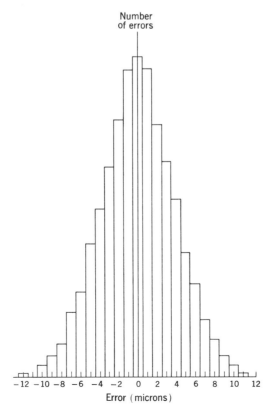

Number of errors

−12 −10 −8 −6 −4 −2 0 2 4 6 8 10 12
Error (microns)

of errors *per unit range of error* (e.g., the number of errors per micron in our example).

A continuous distribution, such as the gaussian destribution of Fig. 2-2, may be interpreted with the aid of calculus. The reader unfamiliar with calculus at this point may skip the details of the proof given in the next paragraph, and simply accept the result, which we now state: the area under a distribution curve between $-x_0$ and x_0 is equal to the number of measurements that would be expected to have errors falling between $-x_0$ and x_0. That area is shown shaded in Fig. 2-2.

According to the previous discussion, a distribution curve is a graph of the number of errors per unit range of error. Expressed in the language of calculus, it is a plot of the function dN/dx versus x. When such a function is integrated between the limits of $-x_0$ and x_0, it yields the number of measurements made with errors falling between $-x_0$ and x_0. This is so because the integral of the derivative of a function is equal to the function itself. We have

$$N_{-x_0 \to x_0} = \int_{-x_0}^{x_0} \frac{dN}{dx} dx$$

Graphically, the integral on the right must be equal to the area under the dN/dx curve between the limits of $-x_0$ and x_0, i.e., to the shaded area in Fig. 2-2.

The particular value of x_0 for which the shaded area is exactly one-half of the total area under the curve is called the *probable error*. It is that value of error which stands a 50:50 chance of being *exceeded* when one additional measurement is performed. In other words, if the distribution curve represents the results of N measurements, the $(N + 1)$st measurement has a 50 percent chance of lying farther than x_0 units away from the mean value in either direction. It, of course, also stands a 50 percent chance of falling within the region between $-x_0$ and x_0.

There is a simple technique for approximating the probable error of a set of experimental data, which comes from the mathematical theory of errors. It involves the computation of the *average deviation* of the set of data. The average deviation is defined as the average *absolute* deviation (or error) of the N measurements from their mean

[2-4] $$\text{a.d.} = \frac{1}{N} \sum_{i=1}^{N} (|m_i - \bar{m}|)$$

For a simple example, let us suppose that the following rather crude set of measurements of the length of a block have been made: 1.0, 1.2, 0.9, 0.8, and 1.1 cm. The average of these data, according to Eq. [2-3], is 1.0 cm (verify this). Then, from Eq. [2-4], the average deviation is

$$\text{a.d.} = \tfrac{1}{5}(|1.0 - 1.0| + |1.2 - 1.0| + |0.9 - 1.0|$$
$$+ |0.8 - 1.0| + |1.1 - 1.0|) = 0.12 \text{ cm}$$

The average deviation is related to the probable error of a normal distribution in a simple way. For large values of N, the statistically valid relation between them is

[2-5]   p.e. $= 0.8453 \times$ a.d.

Thus, the actual probable error of a large collection of data may be seen to be about 15 percent different from the average deviation calculated for that set of data. We will adopt the practice of approximating the probable error by the average deviation calculated by means of Eq. [2-4] for a limited collection of experimental measurements. The factor 0.85 or 0.8543 may be applied, if enough measurements are involved.

Propagation of Errors

We have just discussed how the probable error of a set of experimental data may be computed approximately. In case it is not feasible to perform many repeated measurements of a particular quantity, an estimate of the uncertainty or probable error of that quantity may have to be inferred from the magnitude of the smallest division of the measuring instrument, or from the variation of a small number of measurements, etc. Even after we have performed all measurements of each physical quantity needed to calculate a desired experimental result (such as the density of a solid object), and we have computed or estimated the probable error in each quantity, the question still remains of how each of these errors influences the accuracy of the final result. In other words, we must still know how errors combine, or *propagate*, in order to calculate the probable error in the answer.

In the discussion that follows, we are going to adopt a very conservative attitude toward the propagation of errors through a mathematical computation. In effect, we are looking for a sort of outside limit to the uncertainty associated with the final calculated experimental result. This is most easily done in the case of addition or subtraction. We assume that all errors combine *additively* no matter what the sign of the quantity being added or subtracted. A simple example will suffice to explain the convention. Assume that we must add the following length measurements in order to calculate a desired result:

$$(10.05 \pm 0.03) \text{ cm} + (1.02 \pm 0.02) \text{ cm}$$

The convention that we are following requires that we report the result as (11.07 ± 0.05) cm; our answer must fall somewhere between 11.02 and 11.12 cm. There are better statistical methods of handling this problem, but this simple procedure will certainly provide us with an outside estimate of uncertainty.

The multiplication and division conventions are a little more complicated and we can best illustrate the procedure by considering a specific case. Let us assume that a physical quantity Z is to be calculated from experimentally determined quantities X and Y by means of the formula

[2-6] $Z = X^n Y$

We also assume that the error in each of the latter quantities has been determined; let us denote these errors by x and y. These errors are known as the *absolute* errors in X and Y. The *fractional*, or *relative*, errors in X and Y are x/X and y/Y. The absolute error in Z is denoted by z; its relative error is z/Z. Our task is the determination of the relative and absolute errors in Z, given the absolute errors x and y.

We begin by rewriting Eq. [2-6] so that it includes the errors in X, Y, and Z:

[2-7] $Z \pm z = (X \pm x)^n (Y \pm y)$

Factoring out X and Y, we have

[2-8] $Z \pm z = X^n Y \left(1 \pm \dfrac{x}{X}\right)^n \left(1 \pm \dfrac{y}{Y}\right)$

Applying the binomial theorem to the right-hand side of Eq. [2-8] and dropping all terms beyond the linear ones, we obtain

[2-9] $Z \pm z = X^n Y \left(1 \pm n\dfrac{x}{X}\right) \left(1 \pm \dfrac{y}{Y}\right)$

Multiplying together the binomials, dropping the small term nxy/XY, and multiplying through by $X^n Y$, we obtain for $Z \pm z$

[2-10] $Z \pm z = X^n Y \pm X^n Y \left(n\dfrac{x}{X} + \dfrac{y}{Y}\right)$

We have considered the result obtained in Eq. [2-10] to be of the form $Z \pm z$; therefore, we may identify the error in Z to be

[2·11] $$z = X^n Y \left(n\frac{x}{X} + \frac{y}{Y} \right)$$

Dividing Eq. [2-11] by $Z = X^n Y$, we find the fractional error in Z

[2·12] $$\frac{z}{Z} = n\frac{x}{X} + \frac{y}{Y}$$

We may interpret the final result, Eq. [2-12], in a simple way. The left-hand side of Eq. [2-12] is just the fractional error in Z. The right-hand side of the equation is the sum of the fractional error in X multiplied by the exponent n and the fractional error in Y. This result may be generalized to apply to all quotients and products. We assume, to err always on the conservative side, that all fractional errors are additive. We then find that the fractional error in a function of the type

[2·13] $$Z = X^n / Y^m$$

is of the form

[2·14] $$\frac{z}{Z} = n\frac{x}{X} + m\frac{y}{Y}$$

Verify this relation. It states that the fractional error in a product-quotient type relationship is equal to the sum of the fractional errors in each of the factors multiplied by its exponent.

Once the fractional error in Z has been determined by means of an equation like [2-12] or [2-14], the absolute error in Z may be computed by multiplying the fractional error in Z by the value computed for Z. In the first of the preceding examples, the latter is equal to $X^n Y$. This may be seen by referring to Eqs. [2-11] and [2-12]. Summarizing the results of this treatment of products and quotients, we arrive at the following procedure:

1 Calculate the fractional error in each of the factors.

2 Multiply each fractional error by its exponent in the original formula for Z. Give each of these terms a positive sign.

3 Add together all of the results of step 2 to find the fractional error in the final result, that is, z/Z.

4 Multiply the result of step 3 by the value computed for Z (by substituting the experimental data into the original formula) to find the absolute error in Z.

Let us consider a numerical example involving this kind of expression. Suppose that we have measured the mass and edge length of a metal cube with the following results:

Mass: (2.70 ± 0.04) g
Length: (0.75 ± 0.01) cm

If we denote the mass of the cube by M, and the edge length by L, the density of the metal, according to Eq. [2-1], must be given by

[2-15] $$D = \frac{M}{L^3}$$

To compute the uncertainty in the density, we must add the relative error in M to the relative error in L multiplied by the exponent 3, and then multiply this sum by the computed density

$$\left(\frac{0.04}{2.70} + 3\frac{0.01}{0.75}\right)\frac{2.70 \text{ g}}{(0.75 \text{ cm})^3} = 0.35 \text{ g/cm}^3$$

The reported density would then be $(6.4_0 \pm 0.3_5)$ g/cm^3 or (6.4 ± 0.4) g/cm^3.

How many decimal places should be reported in an experimental result is determined by means of a calculation similar to the one that we have just performed. In that example the uncertainty in the computed density was found to be 0.35 g/cm^3. This means that there is an uncertainty of at least three units in the tenths place in the answer. What then is the significance of a number reported in the hundredths place? Certainly there is no significance at all in a number calculated for the thousandths place. By simply dividing 2.70 g by $(0.75 \text{ cm})^3$, any number of decimal places may be computed for the answer, e.g., $6.400000000 \cdots$ g/cm^3. However, our error analysis tells us that no more than one, or at most two, decimal places have any significance.

4. PROGRAMMING In this section a computer program that will perform error analyses similar to the one just discussed for the metal cube will be described. The program will be required to compute the relative and absolute errors (uncertainties) in the densities of three regular, homogeneous solid objects, along with the density of the material of which each is composed. The three objects, which are in the

FIG. 2-4
Dimensions of solid
objects whose densities
are to be measured

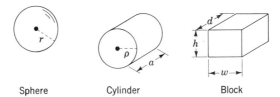

Sphere   Cylinder   Block

FIG. 2-4
Dimensions of solid
objects whose densities
are to be measured

form of a sphere, a cylinder, and a rectangular parallelepiped (block), are illustrated in Fig. 2-4. The sphere has a radius r. The radius and length of the cylinder are indicated by ρ (rho) and a. The three dimensions of the block are w, h, and d.

The computer program first sees to it that dimensions of each object are read in and assigned to appropriate Fortran variables. It must then perform the mathematical operations to compute the density, relative error in density, and absolute error in density for each of the solid objects. Finally, in order to make its computed results available to the experimeter in a useful fashion, it must cause the results to be printed out in tabular form.

The sequence of program steps that accomplish these operations may be conveniently represented in a diagrammatic outline form known as a *flow chart*. The flow chart is a block diagram that illustrates the computer operations effected by the Fortran program in the order in which they are performed. A flow chart is a valuable guide for visualizing the course of a computer program, much as a road map is a guide to the progress made in completing an automobile journey. This is especially true in complicated programs in which there is much *branching*, i.e., in which a number of alternative paths are possible at various points in the program. Flow charts will be presented in each of the following experiments. It will be the job of the reader to interpret them and write appropriate programs to accomplish the desired computations.

The notation that will be used in all flow charts in this book is summarized in Fig. 2-5. It is a fairly standard notation, although a number of minor variations will be found in the literature.[1] In this experiment, we will be primarily concerned with input-output and arithmetic operations, which are indicated by oval and rectangular-shaped program blocks. Each block corresponds to one or more Fortran statements; it indicates the computer operations to be performed in that segment of the program. The program steps are usually written in abbreviated form inside the blocks. We now proceed to discuss a Fortran program that will perform the desired error analysis.

[1] The flow chart symbols that are presented here are the same as those employed in R. V. Jamison, "Fortran Programming," McGraw-Hill Book Company, New York, 1966.

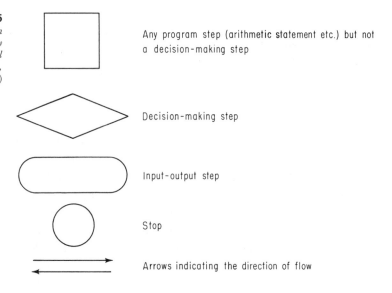

FIG. 2-5
Flow-chart notation (*from "Fortran Programming," by R. V. Jamison, McGraw-Hill Book Company, New York, 1966, with permission.*)

Any program step (arithmetic statement etc.) but not a decision-making step

Decision-making step

Input-output step

Stop

Arrows indicating the direction of flow

A Fortran program to compute the densities of three regular solid objects along with the uncertainty in the density of each of them is presented in Fig. 2-6. The corresponding flow chart is given in Fig. 2-7. The program begins by reading in the dimensions and mass of each object and the values of the errors associated with each of these quantities. Each of the numerical values read in is assigned to a Fortran variable which is named after the quantity it represents.

For example, the Fortran variables R, WS, ER, and EWS in the program of Fig. 2-6 all relate to the solid sphere. Its radius is represented by the Fortran variable R, which corresponds to

FIG. 2-6
Fortran program that computes the densities of objects illustrated in Fig. 2-4 and performs an error analysis of the errors of measurement

```
C       READ IN DATA
        READ (5,100) R, RHO, A, W, H, D, WS, WC, WB, ER, ERHO, EA,
     1  EW, EH, ED, EWS, EWC, EWB
C       COMPUTE DENSITIES
        DS = WS/(1.3333*3.14159*R**3)
        DC = WC/(3.14159*RHO**2*A)
        DB = WB/(H*W*D)
C       COMPUTE RELATIVE ERRORS
        RR = ER/R
        RRHO = ERHO/RHO
        RA = EA/A
        RW = EW/W
        RH = EH/H
        RD = ED/D
        RWS = EWS/WS
        RWC = EWC/WC
        RWB = EWB/WB
C       COMPUTE REL. ERRORS IN DENSITIES
        RDS = 3.0*RR + RWS
        RDC = 2.0*RRHO + RA + RWC
        RDB = RW + RH + RD + RWC
C       COMPUTE DENSITY ERRORS
        EDS = RDS*DS
        EDC = RDC*DC
        EDB = RDB*DB
        WRITE (6,100) DS, DC, DB, RDS, RDC, RDB, EDS, EDC, EDB
100     FORMAT (3F15.5)
        END
```

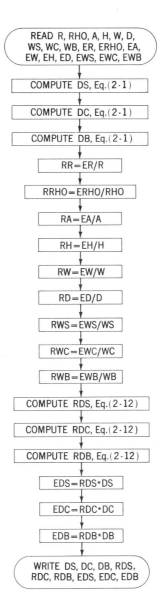

FIG. 2-7
Flow chart describing the
program of Fig. 2-6

READ R, RHO, A, H, W, D, WS, WC, WB, ER, ERHO, EA, EW, EH, ED, EWS, EWC, EWB

COMPUTE DS, Eq.(2-1)

COMPUTE DC, Eq.(2-1)

COMPUTE DB, Eq.(2-1)

RR = ER/R

RRHO = ERHO/RHO

RA = EA/A

RH = EH/H

RW = EW/W

RD = ED/D

RWS = EWS/WS

RWC = EWC/WC

RWB = EWB/WB

COMPUTE RDS, Eq.(2-12)

COMPUTE RDC, Eq.(2-12)

COMPUTE RDB, Eq.(2-12)

EDS = RDS*DS

EDC = RDC*DC

EDB = RDB*DB

WRITE DS, DC, DB, RDS, RDC, RDB, EDS, EDC, EDB

the dimension r in Fig. 2-4. The uncertainty in r is calculated by taking the average deviation of a number of repeated measurements of the radius of the sphere. The uncertainty, or *error*, in r is the value assigned to the variable ER. The weight (actually, the *mass*) of the sphere is associated with the Fortran variable WS. Its weight (mass) error is represented in the program by EWS. The other variables assigned to the input data follow a similar nomenclature. The reason for using WS rather than, say, MS to indicate the mass of the sphere is that MS begins with the letter M, and therefore MS is recognized by Fortran as a *fixed-point* variable. Since all of the calculations in this experiment are performed in *floating point*, a floating-point variable, such as WS, is needed.

The arithmetic portion of the program consists of statements that compute the densities of the three objects (DS, DC, and DB), relative errors in their dimensions and masses (RR, RRHO, RA, RW, RH, RD, RWS, RWC, and RWB) and in their densities (RDS, RDC, and RDB), and finally the *absolute* errors in the densities (EDS, EDC, and EDB). The corresponding arithmetic steps are indicated in the program of Fig. 2-6 and in the appropriate blocks of the flow chart of Fig. 2-7.

The program ends with the printout of the calculated densities, and their relative and absolute errors. It would be possible, of course, to include any or all of the input data in the printout, if it were considered worthwhile to do so. The only modification of the program required to accomplish this would be the addition of the names of the variables that represent these data to the list of variables in the WRITE statement in Fig. 2-6.

The single FORMAT statement in the program, numbered 100, is used both with input and output statements (READ and WRITE). It causes floating-point numbers in 15-column-wide fields to be read in and printed out.

5. PROCEDURE   **a**   Measure the dimensions of each of the three regular solid objects indicated in Fig. 2-4. Use *either* a vernier or micrometer caliper for these measurements. Repeat each measurement several times and determine the average deviation and mean value of each dimension. Record all values.

b   Weigh each of the objects on *either* a platform balance or an analytical balance. Repeat each weight determination several times. Determine the mean value and average deviation of the mass of each solid object. Record these values.

c   Punch the program of Fig. 2-6 on 80-column statement cards, as explained in Experiment 1. Enter the data obtained in parts *a* and *b* on data cards. According to FORMAT statement 100, there should be three data entries on each data card. Each of these numbers is to be punched in floating-point form. It may be located anywhere in its 15-column field, provided it is punched with a decimal point.

d   Have the program run. When all errors have been removed from the program deck and satisfactory answers are obtained, report the final results for the three densities in the form $D \pm d$. Try to identify the solids.

e A suggested variation of the experiment involves the repetition of all measurements with *both* calipers *and* with *both* balances. The Fortran program should perform error analyses with both sets of data. Compare the accuracies of results obtained with the two sets of measuring instruments. For best contrast, use vernier and platform balance together as one measuring combination, analytical balance and micrometer as the other. How do the uncertainties obtained with these two sets of instruments compare?

1. INTRODUCTION Two or more forces acting at a point on a body produce the same effect on it as a single force which acts at the point. That single force which is equivalent to the several individual forces is known as their *resultant*. It is calculated by the mathematical procedure known as *vector addition*. The simplest graphical procedure for accomplishing the addition of two vectors is the parallelogram construction. It is described below.

The parallelogram method of vector addition is illustrated by the practical situation pictured in Fig. 3-1. An 80-lb eastward force and a 60-lb northward-directed force are shown acting on a stone lying on the ground. The common point of application of the two forces is near the center of the stone. They are drawn from that point in the figure. In order to find their resultant, a parallelogram is constructed from them in the following manner. A line is drawn through the tip of each vector parallel to the other. The *magnitude* of the resultant of the two force vectors is equal to the length of the indicated diagonal of the parallelogram constructed in this manner. In this case, because the two forces are at right angles to one another, the parallelogram is a rectangle. The *direction* of the resultant force vector is from the point of application of the forces to the opposite corner of the rectangle. The resultant is labeled **R** in the figure.

The magnitude of the resultant may be found with the aid of the pythagorean theorem. Since the diagonal divides the rectangle into two right triangles, each of whose hypotenuse is R, the magnitude of R may be written

$$R = \sqrt{(60 \text{ lb})^2 + (80 \text{ lb})^2} = 100 \text{ lb}$$

FIG. 3-1
Illustrating the parallelo-
gram method of addition
of two vectors

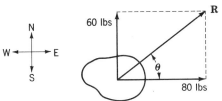

The angle θ between the resultant and the 80-lb vector may be determined trigonometrically. From the definition of the tangent of an angle as the side opposite the angle θ to the side adjacent to θ in a right triangle, we have

$$\theta = \tan^{-1}\tfrac{60}{80} = \tan^{-1}0.75 = 36.9°$$

We may now completely describe the resultant vector **R** as a vector whose magnitude is 100 lb and whose direction is 36.9° north of east.

This process may be extended to any number of vectors which are parallel to the north-south and east-west directions. All north-south vectors are added together algebraically, as are those parallel to the east-west direction. As a result of these additions, there are two vectors, which point north or south and east or west. They are then added together in the same manner as the two vectors in the above example to produce the final resultant.

When several vectors which are to be added together do not make right angles with one another, other methods must be employed to perform the required vector addition. The most straightforward of these methods involves the resolution of all vectors into rectangular components. The process of resolution of vectors is very closely related to vector addition. The rectangular resolution method is discussed in Section 3.

An alternative, trigonometric method of adding three or more vectors is described in Appendix C. The method makes use of the laws of sines and cosines. The vector addition program of this experiment is based on the method of rectangular resolution. The program is discussed in Section 4.

2. APPARATUS   The force table itself is a circular, horizontal metal table with angular-degree marks inscribed in its surface around the circumference of the circle. The force table apparatus is illustrated in Fig. 3-3. Small, light pulleys are fastened to the outer edge of the metal table at desired angles. Strings tied to small weight hangers are passed over the pulleys and attached to a metal ring. The ring fits over a thin supporting rod located at the center of the table in a cylindrical hole. The rod prevents the ring from sliding across the table when the system is not in balance.

When the weight hangers are correctly loaded and are properly located around the circumference of the force table, there is no tendency for the ring to move from the center of the table. The system is then said to be in *mechanical equilibrium*. When the system is in equilibrium, the vector sum of all the weight forces which act on the ring is zero. The support rod may then be removed, and the system will remain in balance.

One method of setting up problems on the force table is to assign weights (forces) and angles of application to all but one of the pulleys. The vector sum of all these weight forces, i.e., their resultant, must be counterbalanced by the weight added to the remaining pulley if equilibrium is to occur. Furthermore, this pulley must be set directly opposite the resultant of the other forces. The correct loading and positioning of the last pulley are done by trial and error.

The magnitude of the resultant may be calculated by adding up all the weights suspended from the last pulley (should the weight of the weight hanger be included?). The total force of these weights constitutes the *antiresultant* of the weight forces from all other pulleys. It is equal in magnitude to their resultant and opposite in direction. When all forces except the last are applied at angles between 0 and 180°, their resultant will be directed at some angle in this range. The antiresultant will then make an angle between 180 and 360° with the x axis.

That the resultant of all but the last force makes an angle between 0 and 180° when these forces are all directed at angles in this range follows from the fact that none of them has a negative y component. Therefore, their resultant may not have a negative y component, either. Hence, it also must make an angle between 0 and 180° with the x axis. These topics are discussed in more detail in Section 3. The restriction that all forces which are to be added be applied at angles less than 180° greatly simplifies the program needed to accomplish their vector addition.

FIG. 3-2
Resolution of a vector v into rectangular components v_x and v_y. The v_x component is negative.

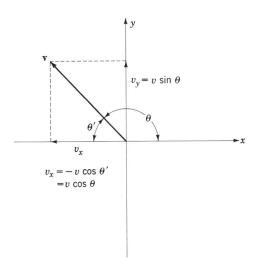

$v_y = v \sin \theta$

v_x

$v_x = -v \cos \theta'$
$\quad = v \cos \theta$

FIG. 3-3
Force table apparatus.
Components of the ap-
paratus include: force
table (T), metal ring (R),
pulleys (P), weight hangers
(H), and support rod (S).

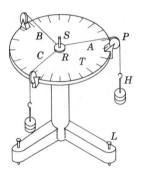

3. THEORY   Vector addition by the method of rectangular resolution is a commonly used procedure. It will be employed in the Fortran program of Section 4 to accomplish the addition of two and three vectors. The method may be summarized in the following steps: (1) resolve each vector into its x and y components, (2) add together algebraically all x components, (3) add together all y components, (4) calculate the resultant by means of the pythagorean theorem, (5) apply the definition of the tangent to find the angle between resultant and x axis. In this section, we will first discuss the resolution of a vector into its x and y components, and then go on to consider the method of vector addition in more detail.

The resolution of a vector into x and y components is illustrated in Fig. 3-2. The vector being resolved is labeled **v**. Its x and y components are v_x and v_y. The process of resolution may be viewed as the inverse of vector addition. Lines parallel to the x and y axes are drawn from the tip of the vector **v** until they intersect the two axes. The x and y components are directed line segments drawn from the origin to the intersection points. A component is taken to be positive if it points along the positive coordinate axis and negative if it points in the opposite direction. The vector **v** has been drawn with a *negative* x component on purpose to show that either sign is correctly handled by the equations used to compute components, Eqs. [3-3] and [3-4] below.

In Fig. 3-2, the angle θ is drawn between the positive x axis and the vector **v**. Its supplement θ' is located between **v** and the negative component v_x. The triangle containing v_x, v_y, and θ' is a right triangle. Therefore, by definition, the side opposite θ' divided by the hypotenuse must be equal to the sine of θ'

[3-1] $$\frac{v_y}{v} = \sin \theta'$$

Rearranging Eq. [3-1], we have

[3-2] $v_y = v \sin \theta'$

Since sines of supplementary angles are equal, Eq. [3-2] may be rewritten as

[3-3] $v_y = v \sin \theta$

By a similar line of reasoning the x component of v may be shown to be given by

[3-4] $v_x = v \cos \theta$

That Eq. [3-4] does provide the correct algebraic sign of the x component is apparent. The quantity v, the *magnitude* of the vector **v**, is positive. Since θ is an obtuse angle, i.e., an angle falling between 90 and 180°, its cosine is negative. Therefore the value of v_x computed by means of Eq. [3-4] must be negative, which is the correct sign for a component that lies along the negative x axis.

In order to perform the addition of any number of vectors by means of the technique of resolution into cartesian components, each vector is first resolved into its components with the aid of Eqs. [3-3] and [3-4]. These components are then added together algebraically as explained in the first paragraph of this section.

The x component of the resultant is found by adding algebraically the individual x components

[3-5] $R_x = \sum_{i=1}^{n} F_{x_i}$

In Eq. [3-5], R_x is the x component of the resultant, and F_{x_i} is the x component of the ith vector being added. The equation for R_y is similar:

[3-6] $R_y = \sum_{i=1}^{n} F_{y_i}$

Once the x and y components of the resultant have been found in this manner, the law of Pythagoras may be employed to calculate the magnitude of **R:**

[3-7] $R = \sqrt{R_x^2 + R_y^2}$

This is the same procedure that was used to calculate the resultant of the 60- and 80-lb forces in the example of Section 1. In fact, once the x and y components of the resultant have been computed, the problem has been reduced to the addition of two perpen-

FIG. 3-4
Three vectors, A, B, and C,
which are to be added. The
angles between these
vectors and the x **axis are**
0°, β**, and** γ**, respectively.**

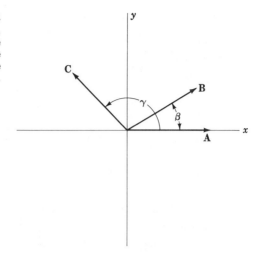

dicular vectors. The same equation may be employed to find the angle ϕ between **R** and the x axis in this case as was used for the addition of two vectors in Section 1:

[3-8] $$\phi = \tan^{-1}\frac{R_y}{R_x}$$

Three vectors **A, B,** and **C** which are to be added together are shown in Fig. 3-4. They make angles of $0°$, β, and γ, respectively, with the x axis of a cartesian coordinate system. For convenience, the vector **A** is taken to lie along the positive x axis. This causes its x component to be equal to its magnitude, and its y component to be zero, thereby simplifying the analysis. It is always possible to choose the coordinate directions in such a way that the x axis is coincident with one of the vectors being added.

4. PROGRAMMING   One of the most convenient features of Fortran is its provision for the evaluation of a standard mathematical function merely by specifying the name of the function and the value of its argument. The Fortran compiler has a *library* of these functions stored in the computer memory in the form of computational routines, usually series approximations. Even with the smallest versions of Fortran, most of the following set of functions is available.

Mathematical function	Fortran notation
$\sin x$	SIN(X)
$\cos x$	COS(X)
$\tan^{-1} x$	ATAN(X)
e^x	EXP(X)
$\ln x$	LOG(X) or ALOG(X)[1]
$\lvert x \rvert$	ABS(X)
$\sqrt{x}$	SQRT(X)

These functions are written in a Fortran statement in the same fashion as they would appear in a mathematical equation. For example, if we wanted to compute the sine of 45°, multiply it by VXO, and store it in the memory area reserved for V, we could write

V = VXO*SIN(0.785398)

Notice that the argument of the sine function is expressed in *radians*. The same is true of the cosine function. The value computed by the arctangent function is supplied in radians also.

It is permissible in Fortran to write expressions for functions of functions. An example is

Y = ALOG(1.05 + ABS(SIN(X + 1.04720)))

The necessity for the three consecutive parentheses at the right of this expression should be verified (i.e., the innermost parenthesis is needed to close the expression for the argument of the sine etc.). This statement computes the natural logarithm of the sum of 1.05 plus the absolute value of sin $(x + 60°)$. It stores the result as y, i.e., the Fortran variable Y.

Fortran Statement Functions   Occasionally, there is need for a mathematical function in a program that is not provided by the particular Fortran compiler being used. An example of such a function which is not generally available in all versions of Fortran is the arcsine. It may be computed with the aid of the arctangent function, which is supplied by most Fortran compilers, by means of the relation

[3-9] $$\sin^{-1} x = \tan^{-1}\left[\frac{x}{(1 - x^2)^{1/2}}\right]$$

---

[1] ALOG(X) is the Fortran IV designation of the natural logarithm. The common logarithm function is ALOG10(X).

The arcsine formula given in Eq. [3-9] may be derived in a straightforward manner. First, let us set $x = \sin \alpha$. Then, by a standard trigonometric identity, $\cos^2 \alpha = 1 - \sin^2 \alpha$, we have

[3-10] $\cos \alpha = (1 - x^2)^{1/2}$

Dividing $\sin \alpha = x$ by Eq. [3-10], we find for the tangent of α

[3-11] $\tan \alpha = \dfrac{x}{(1 - x^2)^{1/2}}$

Taking the inverse of Eq. [3-11] and setting it equal to the inverse of $x = \sin \alpha$, we obtain the desired result, i.e., Eq. [3-9].

When a particular function is required a number of times in a program and it is not supplied by the Fortran compiler, it is convenient to add it to the library of functions that are available for use in that program. This may be done with the aid of an *arithmetic statement function*. The statement function is in the form of an arithmetic statement. It must precede any *executable* statement in the program. The left side of the statement function specifies the name of the function (which must consist of six characters or less, the first being a letter) and includes a *dummy argument*. This argument also appears on the right-hand side of the statement, where it indicates how the function is to be computed. It is permissible to define in this manner functions of more than one argument.[1]

The computation of an arcsine function from the arctangent by means of Eq. [3-9] provides a good illustration of the use of an arithmetic statement function. This kind of statement function would facilitate the programming labor considerably in the case of a rather long program containing many arcsine calculations, if the available Fortran compiler could only supply an arctangent function.

Let us name the arcsine function ARCSIN and designate its dummy index by Y. A suitable statement function with which to compute the arcsine in terms of the arctangent using Eq. [3-9] is

```
ARCSIN(Y) = ATAN(Y/SQRT(1.0 − Y**2))
```

As mentioned above, this statement must precede any executable statement, such as a READ statement, in the program. There-

---

[1] A more complete discussion of the arithmetic statement function will be found in R. V. Jamison, "Fortran Programming," p. 24, McGraw-Hill Book Company, New York, 1966, or D. D. McCracken, "A Guide to Fortran IV Programming," p. 107, John Wiley & Sons, Inc., New York, 1965.

after, whenever the arcsine of any variable or any constant is desired, it may be computed simply by specifying it as the argument of the ARCSIN function. Its arcsine will then be supplied in radians when the program is executed.

As an example of the use of the ARCSIN function, let us suppose that we know the magnitude and y component of a vector **v** and that we would like to know the angle between **v** and the x axis. We may compute that angle (call it θ) with the aid of Eq. [3-3], solved explicity for θ

$$\theta = \sin^{-1}\frac{v_y}{v}$$

The corresponding Fortran arithmetic statement would be

THETA = 57.29578*ARCSIN(VY/V)

The factor 57.29578 has been included to convert the angle θ from radians to degrees.

FIG. 3-5
Flow chart of a program to add the vectors A, B, and C analytically by the method of resolution into rectangular components

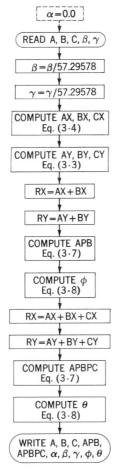

Figure 3-5 is a flow chart of a program to compute the resultant of the two vectors **A** and **B** and of the three vectors **A, B,** and **C**. The program begins by reading in values of the magnitudes of the vectors **A, B,** and **C** and the angles β and γ made by **B** and **C** with the x axis. Next, arithmetic steps in the program carry out the rectangular resolution procedure described in Section 3. After all computations have been performed, the input data and computed results are printed out.

Fortran variables that represent the angles made by the vectors **B** and **C** with the x axis in the program are named BETA and GAMMA. The magnitudes of the resultants of **A** and **B** and of **A, B,** and **C** are assigned to variables called APB and APBPC (standing for A-plus-B, A-plus-B-plus-C). Angles between these resultants and the x axis are represented by PHI and THETA. For completeness, a variable ALPHA, corresponding to the angle between the vector **A** and the x axis, may be defined and set equal to zero. It then may be printed out with β and γ. When this is done, each vector magnitude will be associated in the printout with the angle between the vector and the x axis. A typical printout is shown in Fig. 3-6.

FIG. 3-6
Output produced by the program of Fig. 3-5. Only numerical results are produced by this program; the literal labels had to be added afterwards.

A	B	C	APB	APBPC
1.0000	1.0000	1.0000	1.9319	1.6734
.0000	30.0000	135.0000	15.0000	46.1668
α	β	γ	ϕ	θ

5. PROCEDURE

a Consult your instructor for one or more force addition problems which you are to solve experimentally and theoretically (by means of a computer program).

b Level the force table with the aid of the leveling screws located at the base of the apparatus.

c Position two pulleys at the angles assigned to vectors **A** and **B** (**A** is always located at 0°). Locate a third pulley temporarily at the 270° orientation.

d Slip the small metal ring over the vertical support rod at the center of the force table. Attach three cords to the ring and pass each cord over one of the pulleys. Attach small weight hangers at the ends of the cords. Load each of the hangers located between 0 and 180° with weights equivalent to forces **A** and **B**.

e Locate approximately the direction of the resultant of **A** and **B** by pulling horizontally on the third cord at various angles between 180° and 360°. When the proper direction (opposite the resultant of **A** and **B**) has been found, the ring will "float" at the center of the table, and it will not press against the support rod.

f Move the third pulley to this angle and lightly fasten it there. Now add weights to the weight hanger suspended from this pulley until an approximate balance is achieved. This will be indicated by the lack of a tendency for the ring to move when the support rod is removed.

g Make final adjustments of the angle and magnitude of the third force. By tapping gently on the table, any small tendency it has to move in any direction may be detected. By adding a small weight, perhaps 1 or 2 g, one can get some idea of how much the load may be changed without upsetting the equilibrium situation appreciably. Determine in this way the uncertainty in the magnitude of the measured resultant force.

h Record the magnitude and direction of this force. Subtract 180° from its angle of application to determine the angle made by the resultant of the forces **A** and **B**.

i Set up force **C** with an additional pulley and weight hanger. Repeat steps *e* through *h*.

j Following the flow chart of Fig. 3-5, write a Fortran program that will perform the vector additions analytically and print out both input data and computed results. Prepare a table showing both theoretical and experimental forces and angles. All data and computed results should appear in it in a well-labeled, neatly tabulated form.

EXPERIMENT 4
FREE FALL

1. INTRODUCTION   A body freely falling near the surface of the earth moves with a nearly constant acceleration, which is independent of the mass of the body to good approximation, and which varies only slightly between observation stations on the earth's surface. The principal factors that cause the gravitational acceleration to deviate slightly from a constant value are air resistance and the rotation of the earth. Apart from these effects, which are ordinarily quite small, the gravitational acceleration is a constant approximately equal to 9.8 m/sec² (32.2 ft/sec²). It is usually represented by the letter g.

The acceleration due to gravity is a very important physical quantity. It is the constant of proportionality that relates weight to mass, and it appears in many equations throughout the subject of mechanics.

The equations that relate the position, velocity, and acceleration of a freely falling body are quite simple in form. They are given in Section 3. To measure the position of a freely falling object at equal time intervals is a relatively easy experimental task. With the aid of the equations of free fall, the position-versus-time data may readily be analyzed to provide a good experimental value of g.

In the determination of g, as in many others, we are faced with the problem of performing repeated computations on a large set of experimental data. This is a job for which the digital computer is particularly well suited. Following steps outlined in a program that we have stored in its memory, the computer supervises the entire analysis, including the bookkeeping chores of how many and which operations to perform and in which order to do them.

Implicit in this procedure is the ability of the computer to test some internally stored *counter*. The counter is an integer variable which is incremented by 1 each time a given computation is repeated. When the counter has reached a prearranged value, the computer transfers control to another part of the program, the printout, for example, and the program continues from that point.

This decision-making capability is one of the most important and useful features of the digital computer. It has in large measure made possible many of the widespread applications of the machine. Section 4 on computer programming will be mainly concerned with details of Fortran decision making and transfer of control.

Another feature of the computer, very useful when working with large arrays of numbers, is its provision for dealing with subscripted variables. It is convenient to be able to store a large string of data in the memory in specific locations which are assigned to, say, $x_1, x_2, x_3, \ldots, x_N$. We can perform any desired computations on any or all of these variables and store the results in any order back in the original locations. We may also store these results in any desired order in a second subscripted array, such as $y_1, y_2, y_3, \ldots, y_N$. Another of our principal aims in this experiment is to gain a familiarity with the use of subscripted variables in Fortran programming.

2. APPARATUS

The usual free-fall apparatus consists basically of a freely falling object, called a *bob,* which causes impressions to be made periodically on an adjacent, vertically oriented plate or tape. The latter most often is a paper tape coated with a thin layer of wax. These impressions might be made electrically by a synchronous 60-cycle spark-timing device, or mechanically by a light stylus attached to the falling body. In one version of the electrical method,[1] a protruding metal rim on the falling bob rides between two vertical wire conductors. The waxed-paper tape rests on one wire between the wire and the rim of the bob. At $\frac{1}{120}$- or $\frac{1}{60}$-sec intervals the spark timer applies a high-voltage pulse across the two wires. A spark travels almost instantaneously from one wire to the other through the rim of the bob, puncturing the tape as it does. It leaves an impression in the form of a small dot at the position of the bob at that instant. Figure 4-1 shows the essential components of this type of free-fall apparatus.

In the mechanical method, a vibrating stylus continuously inscribes a sinusoidal trace on the waxed tape as the object to which it is attached falls. The sinusoid is not uniform, of course, due to the acceleration of the bob. The distance between successive maxima grows larger with time. The stylus might be electrically

[1] The Cenco-Behr free-fall apparatus, Central Scientific Company, Chicago, makes use of an electric discharge to produce spark impressions in this manner on a waxed-paper tape.

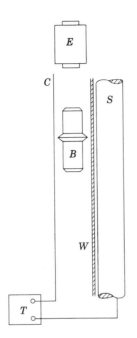

FIG. 4-1
Free-fall apparatus, electrical type. Components of the apparatus include: falling bob (B), waxed-paper tape (W), support column (S), holding electromagnet (E), spark timer (T), and vertical high-tension conductor (C). (*After the Cenco-Behr free fall apparatus, with permission of Central Scientific Company.*)

driven,[1] in which case the frequency of vibration is 60 Hz, or it might be attached to one prong of a falling tuning fork,[2] which then determines the frequency.

The completed tape is spread flat on a smooth table, and the positions of selected spark dots or sinusoid maxima (usually every other one, or every third or fourth one, so that the interval between impressions corresponds to about $\frac{1}{30}$ sec) are measured relative to one of the early dots or maxima. This dot (or maximum) is taken as the origin. Values of the instantaneous velocity and acceleration of the bob are calculated by means of a differences method which is explained in the next section. These computations are performed with the aid of a computer program. It should be written so as to cause all position, velocity, and acceleration-versus-time data to be printed out in a neat tabular form.

An alternative experimental procedure may be employed to determine the gravitational constant. It involves the use of a linear air track. The air-track apparatus and the method of measuring g are both described in detail in Appendix B. The computation of g from the experimental data, which also are recorded on a

[1] This type of electrically driven stylus is found in the Welch free-fall apparatus, Sargent-Welch Scientific Company, Skokie, Ill.

[2] A falling tuning fork apparatus of this kind is described in detail in No. 123 of the Cenco "Selective Experiments in Physics," Central Scientific Company, Chicago.

waxed-paper tape, is exactly the same as the data analysis performed in this experiment. The program needed to compute g from the air-track data is identical to the program described below in Section 4.

3. THEORY

In this section we will derive general expressions for the velocity and acceleration of the bob in terms of its position. These will be useful in the computer analysis of our free-fall data. Before we do so, however, it will be instructive if we illustrate the method of analysis using the specific experimental position data shown in Fig. 4-2. The analysis is outlined in Table 4-1. These data correspond to a $\frac{1}{30}$-sec time interval between spark dots.

Differences are taken between successive position measurements. These differences represent values of the average velocity during each interval in units of cm/interval. They are tabulated in column 3 of the table. When these velocities are multiplied by the conversion factor 30 intervals/sec, they are converted to units of cm/sec. Velocities in cm/sec are given in column 4. We prove below that each of these values of average velocity is also the *instantaneous* velocity at the *middle* of the corresponding interval. For example, the first velocity entry in Table 4-1, 2.92 cm/int. or 87.6 cm/sec, is the instantaneous velocity of the bob at

FIG. 4-2
The falling bob produces spark impressions like the ones shown here (which correspond to the data in Table 4-1). (*After the Cenco-Behr free fall apparatus, with permission of Central Scientific Company.*)

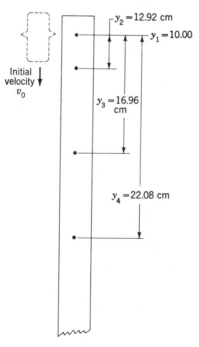

TABLE 4-1
Analysis of Free-fall Data*

Time sec	Position cm	Velocity		Acceleration	
		cm/interval	cm/sec	cm/interval^2	cm/sec^2
0.0000	10.00				
		2.92	87.6		
0.0333	12.92			1.12	1008
		4.04	121.2		
0.0667	16.96			1.08	972
		5.12	153.6		
0.1000	22.08			1.08	972
		6.20	186.0		
0.1333	28.28			1.09	981
		7.29	218.7		
0.1667	35.57				

*After data table on p. 35 of L. Ingersoll, M. Martin, and T. Rouse, "Experiments in Physics," 6th ed., McGraw-Hill Book Company, New York, 1953.

$t = 0.0167$ sec. The next entry is its velocity at $t = 0.0500$ sec, etc.

The average acceleration of the bob in an interval is obtained by subtracting its instantaneous velocity at the beginning of the interval from its velocity at the end of the interval. These acceleration values are tabulated in units of cm/int.^{2} and cm/sec^2 in columns 5 and 6 of Table 4-1. Each one is associated with the instantaneous acceleration at the middle of the corresponding interval. Thus, the first value of acceleration, 1.12 cm/int.^{2} or 1008 cm/sec^2, is taken at $t = 0.0333$ sec.

We now proceed to derive the general expressions for the velocity and acceleration associated in the ith time interval. The origin from which all positions are measured is located at one of the early spark dots or sinusoid maxima. Since the bob will have already fallen some distance before it produces this impression, it will have an initial downward velocity v_0 as it passes the origin. This situation is illustrated in Fig. 4-2. The usual conventions for the coordinates are used, except that the positive y axis is assumed to point downward. The initial position and velocity relative to this coordinate system are

[4-1] $y(0) = 0$

and

[4-2] $v(0) = v_0$

The velocity and position of the bob at any time $t = 0$ are given by

[4-3] $v(t) = v_0 + gt$

and

[4-4] $y(t) = v_0 t + \frac{1}{2}gt^2$

These equations, which are derived in most of the general physics references listed in the bibliography, are consistent with initial conditions [4-1] and [4-2].

Our problem is the experimental determination of the instantaneous acceleration of the falling bob from measurements of its instantaneous position at consecutive, equally spaced instants of time: $t_1, t_2, t_3, \ldots, t_N$. Let us denote the position of the bob at these times by $y_1, y_2, y_3, \ldots, y_N$, respectively. A set of representative y-versus-t data is presented in the first two columns of Table 4-1. These data are plotted in Fig. 4-3.

In principle, we could find the velocity at any time by drawing the tangent to the y-versus-t curve at that time and calculating its slope. However, there is a simpler method that involves only the calculation of differences of experimental data. The theoretical basis of this method follows from the definition of the average velocity $\bar{v}$ during the time interval between t_i and t_{i+1}

[4-5] $\bar{v} = \frac{1}{2}(v_{\text{initial}} + v_{\text{final}})$

The initial velocity during the interval is $v_i = v(t_i)$. The final velocity is $v_{i+1} = v(t_i + \delta)$. When $v(t_i)$ and $v(t_i + \delta)$ are evaluated by means of the constant-acceleration formula of Eq. [4-3] and

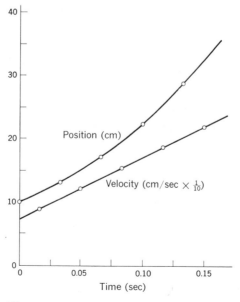

FIG. 4-3
Graphs of bob position and velocity versus time

are substituted into Eq. [4-5], $\overline{v}$ is found to be

[4-6] $\overline{v} = v_0 + g(t_i + \frac{1}{2}\delta)$

Equation [4-6] has a simple interpretation. From Eq. [4-3], the right side of [4-6] must be equal to the *instantaneous* velocity of the bob at $t_i + \frac{1}{2}\delta$, that is, at the *center* of the time interval. Generalizing this result to all time intervals, we have established the fact that *the average velocity during an interval is equal to the instantaneous velocity at the middle of the interval.*

Summarizing the results of the preceding paragraphs, we have for the velocity at the middle of the time interval between t_i and $t_i + \delta$

[4-7] $v(t_i + \frac{1}{2}\delta) = \dfrac{y(t_i + \delta) - y(t_i)}{\delta}$

In order to simplify the notation in Eq. [4-7], let us denote $y(t_i)$ by y_i and $y(t_i + \delta)$ by y_{i+1}. Further, let us represent the velocity at the center of the interval, $v(t_i + \frac{1}{2}\delta)$, simply as v_i. Equation [4-7] may then be written

[4-8] $v_i = \dfrac{y_{i+1} - y_i}{\delta}$

A similar line of reasoning allows us to compute the instantaneous acceleration from the velocities at the *centers* of the intervals just prior to and just after the time at which the acceleration is computed

[4-9] $a(t_i) = \dfrac{v(t_i + \frac{1}{2}\delta) - v(t_i - \frac{1}{2}\delta)}{\delta}$

We may simplify Eq. [4-9] with the aid of the velocity notation used in Eq. [4-8], that is, $v(t_i + \frac{1}{2}\delta) = v_i$ and $v(t_i - \frac{1}{2}\delta) = v_{i-1}$. Setting $a(t_i) = a_i$, we have

[4-10] $a_i = \dfrac{v_i - v_{i-1}}{\delta}$

4. PROGRAMMING   Because subscripts are not easily produced by a typewriter, card punch, or line printer, Fortran employs a subscript notation that can be written on a single line. Parentheses are used to enclose the subscript. This is the same kind of notation customarily used in mathematics for functions of a *continuous* variable. A subscripted variable is really a function of an integer variable, the subscript, which may assume only integer values. There is no reason, therefore, why the subscript should not be written on the same line as the subscripted variable. Thus, the one-dimensional subscripted mathematical variable x_i would be written in Fortran as the subscripted variable X(I). The *argument* of the subscripted variable *is* the subscript.

The subscript itself may be an integer constant, an integer variable, or a simple fixed-point expression, such as $2*J-1$. In order to be a valid subscript, it must at all times be equal to one of the positive integers. Zero and negative integers are not permissible as Fortran subscripts. If a subscript ever assumes a value of zero or a negative integer, Fortran will reject it, and an error message will be the result. V(2) is a valid subscripted Fortran variable, which corresponds to the mathematical variable v_2. V(0), however, is not allowed. $V(2*J-1)$ is a valid subscripted variable provided the value of the integer variable J is 1 or greater.

Dimension, Real and Integer
Before any subscripted variable appears in any program statement, it must previously have been described in a nonexecutable specification or type statement, such as the DIMENSION statement. The DIMENSION informs the computer of the maximum number of elements which the given subscripted variable might contain. The computer then reserves that amount of memory space for the variable. More than one variable may be specified in a single DIMENSION statement. For instance, the three variables A, B, and C may all be defined as one-dimensional floating-point arrays with up to 20 elements by means of the statement

DIMENSION A(20),B(20),C(20)

The type statements INTEGER and REAL may be used in Fortran IV in place of the DIMENSION to specify the size and type of arrays. A REAL statement that will produce the same result as the DIMENSION statement written above is

REAL A(20),B(20),C(20)

The REAL statement describes one or more floating-point variables listed after the word REAL. It is frequently employed for converting variables whose names begin with the letters I through N, which would otherwise be treated as fixed-point variables, to floating point form. Thus, the REAL statement

REAL LENGTH(10), MASS

defines a 10-element one-dimensional array named LENGTH and

an unsubscripted variable named MASS. Both will be treated by the computer as floating-point variables.

The INTEGER statement performs an analogous function for fixed-point variables to that carried out by the REAL statement for floating-point variables. It specifies the sizes of integer-type arrays, and it converts variables whose names begin with letters A through H and O through Z into fixed-point form. A typical INTEGER statement is

INTEGER N(10),S2(10),DIGIT

It defines two 10-element fixed-point arrays named N and S2 and an unsubscripted fixed-point variable named DIGIT.

Not all of the elements specified by a DIMENSION, REAL, or INTEGER statement need actually be assigned values during the execution of a program. For instance, numerical values might only be read into the first six elements of the integer variable N defined by the preceding INTEGER statement. Then only these elements (or only a portion of them) might be operated upon when the program is executed. The remaining four elements of N specified in the INTEGER statement would not be used.

Transfer of Control

Ordinarily, the statements of a Fortran program are executed in sequential order, one statement after another, until all of its statements have been performed. It is important to be able to interrupt the consecutive order of execution of Fortran statements and jump forward or backward in a program. A statement that causes a break in the sequential *flow* of a program is known as a *control statement*. The process of jumping forward or backward is referred to as *transfer of control* or *branching*.

An important and frequently used control statement is the IF statement. It provides both Fortran II and Fortran IV with a "decision-making" capability. On the basis of the current value of some variable or expression that it *tests* (compares with zero), it may cause a program to branch in one of two or three different directions. The IF statement is treated in detail later in this section.

One particular kind of branching with numerous useful applications in computer programming is called *looping*. It is the process of transferring control back to some earlier point in a program and repeating all intervening steps for different values of various Fortran variables or data. The statements included between the statement to which control is transferred and the point at which the branching occurs are said to constitute a *loop*.

A Free-Fall Data Analysis Program

Figure 4-5 is the flow chart of a Fortran program that will perform the analysis of the experimental y-versus-t data obtained in a free-fall experiment. It will produce a tabulated print out similar to Table 4-1.

The program starts out by reading in values of N, the total number of measured spark dots, and DELTA (or DELT), the time interval between successive spark dots. The program then causes values of spark positions to be read into the first N elements of the subscripted variable Y. This is accomplished in a counting *loop*, which is discussed later in this section.

Before the second loop in Fig. 4-5 begins, the variables T and V must be initialized, and the subscript I must be set equal to 2. T(1) is equated to zero. V(1) is computed from Y(1) and Y(2) by means of Eq. [4-8],

V(1) = (Y(2) − Y(1))/DELT

As explained in Section 3, V(1) represents the instantaneous velocity of the bob at $t = \frac{1}{2}\delta$, *not* at $t = 0$. Y(1) is its position at $t = 0$.

After the subscript I has been equated to 2, the computational loop is started. It performs the analysis of the spark-tape data by means of Fortran arithmetic statements which are equivalent to Eqs. [4-8] and [4-10]. The subscript I is incremented by one each time the loop is traversed. The computation continues until T(N−1), V(N−1), and A(N−1) have been computed. T(N) is calculated separately *after* the loop has been completed. It is not possible to compute V(N) since there is no Y(N+1). Therefore, there is no A(N) either.

The data are printed out in another counting loop. Before the loop is started, T(1) and Y(1) are printed out on one line, and V(1) is printed below them on a second line. The subscript I is set equal to 2, and the printout loop is begun. It continues until the V(N−1) entry has been printed. Finally, T(N) and Y(N) are printed out.

The data table is now complete. It should look like Table 4-1. This kind of printout is made possible by printing T(I), Y(I), and A(I) on one line under appropriate F specifications so that a blank field is left where V(I) will appear on the next line. This could be accomplished, for example, by the following set of output statements

```
  WRITE (6,1) T(I), Y(I), A(I)
1 FORMAT (2F10.3,F20.3)
  WRITE (6,2) V(I)
2 FORMAT (F30.3)
```

Take one row of T, Y, V, and A data from Table 4-1 and write the exact output that would be produced from it by the program segment above.

The IF Statement and Looping

The new feature in the flow chart of Fig. 4-5 is the diamond-shaped decision block. It implies a transfer of control based upon the results of a test of the sign of some expression. The basic Fortran decision-making statement (in both Fortran II and IV) used to accomplish this transfer of control is the IF statement. The IF has the general form

IF (expression) A, B, C

The expression inside the parentheses in the preceding IF statement may be any valid integer- or floating-point expression. A, B, and C are statement numbers which identify three statements in the program. If the value of the enclosed expression at the time the IF statement is executed is negative, control is transferred to A; if zero, to B; and if positive, to C.

An important use of the decision-making process is the testing of some integer *counter* in a programming loop. Let the counter be named I and let its maximum value be specified by the integer variable N. The value of N must be read in before the counter is incremented and the loop which it controls is started. I is then set equal to 1 and the loop calculations or operations are initiated. After one pass through the loop, the counter is incremented by the fixed-point arithmetic statement

I = I + 1

At this point, the counter must be tested to see if it is greater than, equal to, or less than N. This may be accomplished by an IF statement such as

IF (I − N) 3, 3, 6

Statement 3 is the first executable statement in the loop. If I is less than or equal to N, control is transferred to statement 3 and the loop is executed one more time. If not, control is transferred to statement 6, which is located outside the loop.

A simple example of a loop of this type is

```
  I = 1
3 READ (5,1) X(I)
1 FORMAT (F10.4)
  I = I + 1
  IF (I − N) 3, 3, 6
6 . . .
```

This loop causes N values of the subscripted variable X to be inputted.

In order to put the idea of a computational loop in a more concrete form, let us consider the evaluation of the terms of the geometric progression

$$a, ar, ar^2, \ldots, ar^{n-1}$$

There are in all n terms in this progression. Each differs from the preceding term by the common ratio r. The flow chart of a program to compute all n terms (numbered $1-n$) is given in Fig. 4-4a. The corresponding program is shown in Fig. 4-4b. The terms of the progression are represented by the elements of the subscripted variable T. The program accomplishes the evaluation and printout of these terms by means of two counting loops of the kind discussed in the previous paragraph, one to perform the evaluation of terms, and the other to effect the printout of results.

5. PROCEDURE   a   The apparatus may be equipped with a holding electromagnet to retain the bob until the time for release. Connect the magnet, located at the top of the apparatus, in series with a low-voltage dc source and a switch. Either two dry cells in series or a single 2-volt storage cell would be suitable power sources.

FIG. 4-4
Program to evaluate the terms of a geometric progression. The corresponding flow chart is given in (a); the program itself in (b).

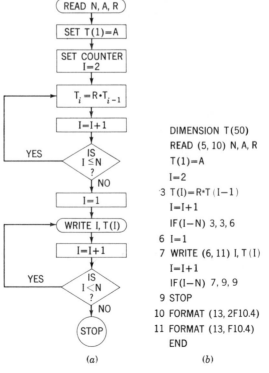

```
DIMENSION T(50)
READ (5, 10) N, A, R
  T(1)=A
  I=2
3 T(I)=R*T (I−1)
  I=I+1
  IF (I−N) 3, 3, 6
6 I=1
7 WRITE (6, 11) I, T (I)
  I=I+1
  IF (I−N) 7, 9, 9
9 STOP
10 FORMAT (13, 2F10.4)
11 FORMAT (13, F10.4)
  END
```

(a)   (b)

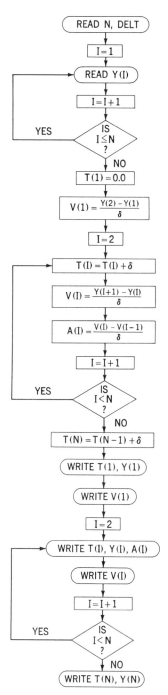

FIG. 4-5
Flow chart of a program to analyze experimental free-fall data

b Carefully level the free-fall apparatus. Precise vertical orientation of the electrical wires or mechanical guide rods is very important in this experiment. The plumb line supplied with the apparatus will aid in the alignment.

c Feed a fresh length of tape from the supply roll and position it alongside the vertical support or guide rods. Make sure that it is in the correct lateral position, so that the bob as it falls will produce impressions down the center of the tape. Perform a similar alignment of the glass plate, if that form of apparatus is used.

d Suspend the bob from its support at the top of the apparatus. Steady it so that it is not swinging appreciably at the time it is released. Turn on the electrical power that drives the stylus or energizes the spark timer. Release the bob. If a spark timer is being used, switch on the high voltage just before the bob is released, and switch it off just after the bob comes to rest at the bottom of the apparatus.

e Remove the completed tape or plate and spread it flat on a table. Fasten it to the table with mending tape after smoothing it to remove any wrinkles (if the waxed tape is used). Lay a 2-m stick on the waxed tape or plate parallel to the trace. It is advisable to rest the meter stick on its narrow edge so that its millimeter division marks are in contact with the trace. It is also a good idea to fasten down the meter stick to avoid any accidental shift during the course of the measurements.

f Align the 10-cm division of the 2-m stick with one of the early spark dots or sinusoid maxima. This point will be taken as the origin. Read the positions of the selected impressions (every other impression or every fourth one, etc., corresponding to about $\frac{1}{30}$ sec time intervals). Estimate positions to the nearest tenth of a millimeter. Note that all positions are to be measured relative to the *origin*. If the air track is used, a longer time interval will be needed because the acceleration is smaller and the trace impressions are closer together. A time interval long enough so that the position measurements are similar to those in Table 4-1 should be chosen.

g Write a computer program that will perform the analysis of the data and print out results in a tabulated form similar to Table 4-1. The flow chart of the computation procedure is indicated in Fig. 4-5. A sample program is presented in Appendix D. The reader ought to try to write his own program without referring to this sample program. The computer printout should be completed by supplying neatly lettered column headings and labels.

h Plot three curves on the same time axis, using three different scales for the ordinates: y versus t, v versus t, and a versus t. The last of these, the a versus t curve, should be plotted with an expanded ordinate scale, and only the range close to 980 cm/sec^2 should appear on the graph. Use different colors for the three curves, or different kinds of lines, such as solid, dashed, and dotted, etc.

i Calculate an average value of the acceleration g from your results.

Compute the average deviation. Report a final value for g including only as many significant figures as you consider justifiable. Include a plus-or-minus estimate of the error in g based upon the average deviation. Report the experimental value of g in the form

(980 ± 2) cm/sec^2

or

(9.80 ± 0.02) m/sec^2

1. INTRODUCTION A body moving at constant speed in a circular path experiences an acceleration directed toward the center of the circle. The magnitude of the acceleration is shown in the general physics references cited in the bibliography to be given by

[5-1] $a = \dfrac{v^2}{r}$

Here, r is the radius of the circle in which the body or particle is moving and v is its speed, the magnitude of its velocity. Because of its direction, the acceleration is termed *centripetal* (from the Latin for *center-seeking*). Figure 5-1 shows a top view of a body of mass m, called a *bob*, traveling in a horizontal circle of radius r with velocity **v**. The centripetal acceleration vector is labeled **a.**

According to Newton's second law of motion, whenever a body of mass m is moving with acceleration **a** in a given direction, it experiences a net force **F** in that direction. This net force, which is the vector sum of all the forces acting on the body, can be written as

[5-2] $\mathbf{F}^{\text{net}} = \displaystyle\sum_{i=1}^{n} \mathbf{F}_i = m\mathbf{a}$

This is the vector statement of the second law.

In particular, when a body or particle of mass m is moving in a circular path at constant speed, and its acceleration is directed

FIG. 5-1
Top view of a body of mass m moving in a circle of radius r. The velocity and acceleration vectors are indicated in the diagram.

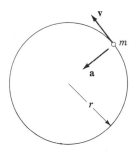

toward the center of the circle, Eq. [5-2] requires that there be a net force in this direction acting on the body. This force is called the *centripetal force*. From Eqs. [5-1] and [5-2], its magnitude must be

[5-3] $$F_c = \frac{mv^2}{r}$$

It is a force which may be exerted by a cord or a rod connecting the body to the center of rotation, or, as in the case of the revolution of a planet around the sun, the force of gravitational attraction which the sun exerts on the orbiting planet.

2. APPARATUS Both motor-driven and manually operated versions of the centripetal force apparatus may be used in this experiment. Almost all commercially available units, whether driven by motor or hand, are similar in operation. A bob of mass m is connected by a horizontal arm of length r to a rotatable vertical shaft. The bob is suspended from the end of the arm, or, in one form of the apparatus, it is attached directly to a stiff spring which takes the place of the horizontal arm. The general form of the centripetal force apparatus is illustrated in Fig. 5-2. A spring is attached to the bob and to the rotation axis in the apparatus shown there.

The drive motor must be of the variable speed type so that the orbital speed of the bob can be adjusted continuously. According to Eq. [5-3], as the speed of the bob increases, so does the centripetal force that is needed to keep it moving in its circular orbit. The spring supplies this force. It is stretched more and more as the speed of the bob is increased; i.e., by Newton's third law, as the spring applies more and more inward centripetal force to the bob, the bob applies more outward, stretching force to the spring. The same situation occurs with the manual rotator.

FIG. 5-2
Centripetal force apparatus. Components are: vertical shaft (V), support arm (A), bob (B), spring (S), and pulley (P).

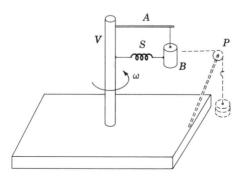

The speed is increased until the bob travels directly beneath the end of the support arm. Experimentally, this is accomplished by mounting a small paper indicator, e.g., a used safety match, at the correct orbit radius. When the bob just begins to brush against the indicator, the proper orbit has been attained. The speed of revolution must be continually adjusted during the actual determination to keep the bob in this orbit.

Some versions of the centripetal-force apparatus include a built-in revolution counter. The counter is provided with some sort of clutch-drive mechanism so that it can be quickly engaged and released by the operator. The counter is engaged at the same instant a clock timer is started; it is disengaged after a prearranged time interval has elapsed. The frequency of revolution is determined by subtracting the initial and final counter readings and dividing by the time interval.

In the case of a manually operated centripetal-force apparatus which does not have provision for a counter, the procedure is modified. A clock timer is started at some convenient point in the orbit. A specified number of revolutions, say 20, are timed, and the frequency is calculated as before, by dividing the number of revolutions by the total time elapsed.

A static technique may be employed to measure directly the centripetal force. With the apparatus at rest, a cord is attached to the bob and passed over a light pulley. It is fastened to a weight hanger at the other end (dashed lines in Fig. 5-2). Weights are added until the spring is pulled out to the same radius it had while the bob was moving in its correct orbit. The force then acting on the spring is

[5-4] $F_{\text{spring}} = m_{\text{weights}} g$

In Eq. [5-4] the mass must include that of the weight hanger; g is the acceleration due to gravity. Since the force exerted by the spring while the bob is in motion is equal to the force calculated in [5-4], this equation may be used to compute F_c directly.

3. THEORY

Equation [5-3] gives the formula for the magnitude of the centripetal force that acts on the bob while in orbit. It involves m, r, and v. Experimentally, we measure m, r, and f, the frequency of revolution. The orbital speed v, however, is simply related to the frequency f. If we let T represent the time of one revolution, i.e., the *period*, then $f = 1/T$. Now, since the distance traveled by the bob in one revolution is the circumference of the orbit, $2\pi r$, the velocity is given by

[5-5] $$v = \frac{2\pi r}{T} = 2\pi rf$$

Combining Eq. [5-5] with Eq. [5-3], we obtain the magnitude of the centripetal force

[5-6] $$F_c = 4\pi^2 mrf^2$$

The usual experimental determination of the frequency of revolution proceeds as follows. The counter is read initially, then it is engaged for a given time interval, read again, and the whole procedure repeated a specified number of times. The average frequency is calculated by summing successive differences of counter readings and dividing by the product of the number of intervals and the time of one interval.

It so happens that an average calculated in the way just described discards all counter readings except the initial reading and the final reading. That this is so is easily demonstrated. Let us denote the initial reading by R_1; the next, R_2; etc. The final counter reading is designated R_n. Thus there are in all n readings. Following the procedure of the preceding paragraph, we form successive differences $(R_2 - R_1)$, $(R_3 - R_2)$, $(R_4 - R_3)$, . . . , add them, and divide by the time interval T multiplied by n:

[5-7] $$\bar{f} = \frac{(R_2 - R_1) + (R_3 - R_2) + \cdots + (R_n - R_{n-1})}{nT}$$

After removing parentheses in Eq. [5-7] and canceling as many terms as possible, we have finally for $\bar{f}$

[5-8] $$\bar{f} = \frac{R_n - R_1}{nT}$$

The result calculated for $\bar{f}$ in Eq. [5-8] disregards all intermediate readings, $R_2, R_3, \ldots, R_{n-1}$. The same answer would be obtained if we were to make only initial and final readings and not any others.

One way of avoiding this difficulty involves the "method of differences."[1] This technique requires that an even number of readings be made. Let us call them $R_1, R_2, \ldots, R_n$. The readings are divided into two groups, R_1 to $R_{1/2n}$ and $R_{1/2n+1}$ to R_n. Differences are taken between corresponding readings in the two

[1] L. Ingersoll, M. Martin, and T. Rouse, "Experiments in Physics," 6th ed., p. 5, McGraw-Hill Book Company, New York, 1953.

groups, e.g., $(R_{1/2n+1} - R_1), (R_{1/2n+2} - R_2)$, etc. These differences are added together, and their sum is divided by the total time represented by the terms of the sum.

A specific example will help to clarify the procedure. Let us assume that we have made eight successive readings of a revolution counter. These readings are divided into the two groups R_1 to R_4 and R_5 to R_8. The sum of differences $(R_5 - R_1) + (R_6 - R_2) + (R_7 - R_3) + (R_8 - R_4)$ is formed. Each term in the sum involves a difference between measurements separated by four time intervals. There are four terms in the sum. Thus, the denominator of the expression for $\bar{f}$ must be equal to 16 time intervals, so $\bar{f}$ is given by

$$\bar{f} = \frac{(R_5 - R_1) + (R_6 - R_2) + (R_7 - R_3) + (R_8 - R_4)}{16T}$$

In this expression, T is the time interval between two successive readings.

When parentheses are removed from the above expression for $\bar{f}$ and terms are rearranged, $\bar{f}$ may be written

$$\bar{f} = \frac{-R_1 + R_2 - R_3 + R_4 - R_5 + R_6 - R_7 + R_8}{16T}$$

No term in the numerator cancels with any other, all intermediate readings are included in the average, and each term is given equal weight in the computation. Therefore, this method provides a simple, efficient way to compute an average that takes into account all readings and gives each of them equal importance in the final result. We note that the alternating finite series in the numerator may be expressed concisely with the aid of the summation notation as

$$\sum_{i=1}^{8} (-1)^n R_i$$

This result should be verified by writing out the explicit terms of the sum.

In the general case, differences are taken between measurements which are $\frac{1}{2}n$ intervals apart. There are in all $\frac{1}{2}n$ such measurements. Therefore, the denominator of the expression for $\bar{f}$ must be equal to the product of $\frac{1}{2}nT$ and $\frac{1}{2}n$, that is, $\frac{1}{4}n^2T$. The general form of the expression for $\bar{f}$ may be expressed by

[5-9] $$\bar{f} = \frac{4}{n^2T} \sum_{i=1}^{1/2n} (R_{1/2n+i} - R_i)$$

We may derive an alternative form of this equation by generalizing to n readings the alternating series that we wrote previously.
This leads to a sum with n rather than $\frac{1}{2}n$ terms

$$[5\text{-}10] \quad \bar{f} = \frac{4}{n^2 T} \sum_{i=1}^{n} (-1)^n R_i$$

As a check on their validity, the two expressions for $\bar{f}$ given in
Eqs. [5-9] and [5-10] should be written out explicitly for the case
of $n = 8$. The resulting expressions should be identical to the one
previously obtained.

4. PROGRAMMING

In this experiment, we will write a program to compute the
average frequency by the method of differences for n readings of
the revolutions counter of a centripetal force apparatus. This
frequency will then be substituted into Eq. [5-6] to compute the
centripetal force which acts on the rotating bob. In the computa-
tion of $\bar{f}$, we must evaluate the sum of either $\frac{1}{2}n$ or n terms, depending on whether Eq. [5-9] or [5-10] is used in the calculation.

Summations in Fortran

Summations are performed in Fortran with the aid of loops of
the kind discussed in Experiment 4. The ith loop must evaluate
the ith term of the sum and add it to the sum of the preceding
$i - 1$ terms. This is accomplished conveniently by making use of
the basic property of the Fortran equals sign. The equals sign
replaces the contents of the storage location assigned to the
variable on its left with the value computed for the expression
on its right. It causes this replacement to take place immediately
after that expression has been evaluated.

As a first example, let us consider the type of counting loop
employed in the preceding experiment. The incrementing (summing) statement used in a loop of this kind has the form

I = I + 1

Each time this statement is executed, I is incremented by 1. If
I initially had been set equal to zero, its value at any time during
the execution of the program would indicate the particular pass
through the loop in progress at the time. The running sum is kept
in the storage location allotted to the variable I.

Another example of Fortran summation is presented in the

program of Fig. 5-3a. This program computes and prints out the sum of the first i terms of the geometric progression

$$a, ar, ar^2, \ldots, ar^i, \ldots, ar^n$$

In this scheme of indexing, the first term is assigned the index $i = 0$. The program accomplishes the summation by computing the ith term from the $(i - 1)$st term by means of the statement

TERM = TERM*R

and adding it to the sum of the preceding $n - 1$ terms by means of the statement

GPSUM = GPSUM + TERM

Each successive term in the progression is found from the preceding term by multiplying it by the common ratio r. This is done by the first of these two statements. The output of this program for $n = 20$ is given in Fig. 5-3b.

As a final illustration of the Fortran summation process, let us consider the computation of the mean (average) of a set of n measurements, $x_1, x_2, \ldots, x_n$. The standard formula for the mean of n measurements is

[5-11]
$$\bar{x} = \frac{1}{n} \sum_{i=1}^{n} x_i$$

We assume that the values obtained for the n measurements have already been read into the first n elements of the subscripted variable X. The summation is then performed by a repeated execution of the Fortran statement

SUMX = SUMX + X(I)

FIG. 5-3
Program to compute the sum of i terms of a geometric progression. (a) The Fortran program; (b) the output produced for $n = 20$.

```
READ (5,10) N, A, R
I = 0
TERM = A
GPSUM = A
5 TERM = R*TERM
GPSUM = GPSUM + TERM
I = I + 1
WRITE (5,10) I, TERM, GPSUM
IF (I - N) 5, 8, 8
8 STOP
10 FORMAT (I7, 2F12.4)
END
```

(a)

	TERM	SUM
1	1.01 00	2.01 00
2	1.02 01	3.03 01
3	1.03 03	4.06 04
4	1.04 06	5.10 10
5	1.05 10	6.15 20
6	1.06 15	7.21 35
7	1.07 21	8.28 57
8	1.08 29	9.35 85
9	1.09 37	10.45 22
10	1.10 46	11.55 68
11	1.11 57	12.68 25
12	1.12 68	13.80 93
13	1.13 81	14.94 74
14	1.14 95	16.09 69
15	1.16 10	17.25 79
16	1.17 26	18.43 04
17	1.18 43	19.61 47
18	1.19 61	20.81 09
19	1.20 81	22.01 90
20	1.22 02	23.23 92

(b)

A short Fortran program which accomplishes the averaging of n measurements is

```
    DIMENSION X(20)
    READ (5,10) N
    I = 1
  4 READ (5,11) X(I)
    I = I + 1
    IF (I − N) 4, 4, 8
  8 SUM = 0.0
    I = 1
 19 SUM = SUM + X(I)
    I = I + 1
    IF (I − N) 19, 19, 21
 21 AVGE = SUM/FLOAT(N)
    WRITE (6,11) AVGE
 10 FORMAT (I3)
 11 FORMAT (F12.4)
    END
```

The program consists of two counting loops, one to read in the elements of the one-dimensional array X, and the other to perform the summation required by Eq. [5-10].

After the x_i's have been read in, the subscript I and the variable SUM are initialized. I is set equal to 1, and SUM is equated to zero before the computational loop is started. The storage location occupied by the variable SUM must be cleared (by initializing SUM to zero) before anything is added to it. We have no way of knowing ordinarily what the contents of this storage location might be as a result of the program previously executed by the computer. If we did not initially clear it, the computer would try to add our x_i's to whatever was left in it. This might not even produce a numerical result, and it very well might cause termination of the program. At least, we would not be very surprised by an incorrect answer if we had omitted this initialization step.

The remainder of the program consists of statements that increment the subscript i, form the sum of the x_i's, test the subscript, and compute the average. The loop is terminated after the $i = n$ term has been added to the sum. The average is assigned to a variable named AVGE. The only new noteworthy feature in the program is the introduction of the FLOAT function in the statement that computes the value of AVGE. The FLOAT function converts its integer argument (N, in this case) into floating-point form. This ensures that all arithmetic operations on the right-hand side of statement 21 will be performed in floating point.

It should be pointed out that the conversion of N to a floating-point variable could be accomplished without the use of the FLOAT function. The conversion could be accomplished by the Fortran statement

21 XN = N

The computation of the mean might then be performed by means of a statement like

AVGE = SUM/XN

When these two statements are substituted for the previous statement 21, they produce precisely the same value of AVGE as it does.

Unconditional Transfer of Control, the GO TO Statement

Many programming situations occur in which it is necessary to transfer from one point in a program to another as soon as the first point is reached. Without performing any test or comparison, the computer shifts control to the second point (statement). This so-called *unconditional transfer of control* is accomplished by means of the GO TO statement. It does precisely what its name indicates, e.g.,

GO TO 15

tells the computer to transfer control *immediately* to statement 15.

One useful application of the unconditional transfer of control is the repetition of an entire program with a different set of input data. This is conveniently accomplished with a GO TO as the last *executable* statement in a program. It is referred to a numbered READ statement near the beginning of the program. The READ causes a new set of data cards (or, perhaps, just a single data card) to be read. After the new data has been inputted, the program is executed a second time etc. This is a kind of looping process similar to the counting loops previously described. However, this loop contains no decision-making statement, and it would continue indefinitely, except for the fact that there are a finite number of data cards. After the last card has been read and the program has been executed, the GO TO sends control back to the READ statement. There are now no more cards to be read, and the program is terminated.

The program to compute the mean value of *n* measurements could be extended to compute the mean of as many sets of data as have been punched on data cards and to print out the mean of each set of data. A modification of the averaging program which would perform these operations is indicated below.

```
    DIMENSION X(20)
  2 READ (5,10) N
    .

    .

 21 AVGE = SUM/FLOAT(N)
    WRITE (6, 11) AVGE
    GO TO 2
 10 FORMAT (I3)
 11 FORMAT (F12.4)
```

We notice that the GO TO statement is not the last statement in the program. It is the last *executable* statement. Each set of data may consist of any number of measurements from 1 to 20. The READ statement numbered 2 causes the value of n, an integer equal to the total number of measurements in the *forthcoming* data set to be read in first. This number is stored as the Fortran variable N. It limits the data input loop to n passes, and it also sets n as a limit to the number of times the summing loop is to be executed.

5. PROCEDURE   **a** Level the centripetal force apparatus. If the manually operated version of the apparatus is being employed, set the support arm from which the bob is suspended to the desired radius. Adjust the counterbalance weight properly. This will have been accomplished when there is no tendency for the support arm to tip in either direction when the screw that fastens it is loosened.

b If the motor-driven rotator form of the apparatus is used, perform the necessary leveling adjustments so that the plane of rotation of the bob is horizontal.

c Start the bob revolving about the vertical rotation axis. Adjust the angular speed of rotation so that the bob revolves in its correct orbit, as discussed in the section on apparatus.

d Follow the procedure discussed in that section for determining the angular velocity of rotation. When using a manually driven rotator, record the times at which the 20th, 40th, 60th, etc., revolution have been completed. The timing stopwatch or stopclock should be started at the *beginning* of the first revolution. Time an odd number of intervals. Define t_1 to be equal to zero, t_2 equal to the time recorded at the end of the 20th revolution, etc. There will then be an even number of t_i's.

e If the motor-driven apparatus is used, record the counter reading before the revolution counter is engaged. Engage the counter. Wait 30 sec; then disengage the counter and record its reading. Repeat this procedure and obtain counter readings corresponding to any *odd* number of 30-sec intervals. This ensures that the number of R_i's is even, as required in the method of differences.

f Stop the apparatus. Measure and record the orbit radius. Determine and record the mass of the bob (which may be stamped on its surface). Attach a cord to the bob and pass it over the pulley to a suspended weight hanger. Add weights sufficient to stretch the spring to the orbit radius. Compute F_c by means of Eq. [5-4].

FIG. 5-4
Flow chart of a program to compute centripetal force using the method of differences.

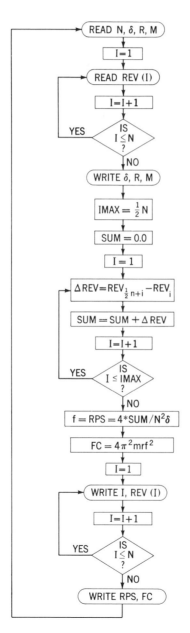

g Refer to the flow chart of Fig. 5-4. Write a computer program that will accomplish the computation of the angular speed of revolution (in revolutions per second) by the method of differences. The program should also provide for the computation of F_c from Eq. [5-6] and the printout of all data and results in tabular form. A representative sample program is included in Appendix D.

EXPERIMENT 6
NEWTON'S SECOND LAW

1. INTRODUCTION The basic principle underlying classical dynamics is Newton's second law of motion. The second law relates the acceleration of a body to the net force that produces it. The mathematical relation between these quantities, according to the second law, is a linear one

[6-1] $\mathbf{F}^{\text{net}} = m\mathbf{a}$

The constant of proportionality in Eq. [6-1], m, is called the *mass* of the body. It may be considered to be a measure of the inertia of the body or of its resistance to a change in its state of motion. The more massive a body is, the greater must be the applied force that imparts a given acceleration to it.

The statement of the second law presented in Eq. [6-1] is a vector equation. The left-hand side of this equation is the net vector force acting on the body. It is equal to the *vector* sum of all forces exerted on the body. If we denote the individual forces by $\mathbf{F}_1, \mathbf{F}_2, \ldots, \mathbf{F}_i, \ldots, \mathbf{F}_n$, we may write Eq. [6-1] in terms of these forces as

[6-2] $\displaystyle\sum_{i=1}^{n} \mathbf{F}_i = m\mathbf{a}$

Equation [6-2] may be written in the form of three component equations in rectangular coordinates

[6-3a] $\displaystyle\sum_{i=1}^{n} F_{x_i} = ma_x$

[6-3b] $\displaystyle\sum_{i=1}^{n} F_{y_i} = ma_y$

[6-3c] $\displaystyle\sum_{i=1}^{n} F_{z_i} = ma_z$

The sum appearing in each of the Eqs. [6-3] is the *algebraic* sum of the force components in the particular coordinate direction specified by that equation. For instance, the first of these equations, Eq. [6-3a], involves the sum of all x components of force acting on the body. Each must be assigned a positive or a negative sign depending upon whether that component points along the positive or negative x axis. The sum of these force components may then be positive or negative. If the sum is positive, Eq. [6-3a] requires that the x component of acceleration also be positive, i.e., that the body have a component of acceleration in the positive x direction.

In the simpler case of one-dimensional motion, only one component equation of the second law is necessary to describe the motion of the body. In this case, any one of Eqs. [6-3] is sufficient to provide a complete description of its motion, and there is no longer any need for subscripts to denote the particular component involved. For the case of one-dimensional motion, the second law may be written

[6-4] $$\sum_{i=1}^{n} F_i = ma \qquad \text{one dimension}$$

It is still necessary to provide each force component in the sum in Eq. [6-4] with its correct sign, i.e., to relate each force component to a positive direction. The acceleration may then have either a positive or a negative sign associated with it. If its sign is positive, it must point along the positive direction, and vice versa.

2. APPARATUS Figure 6-1 illustrates the basic features of the apparatus that will be used to study Newton's second law in this experiment. There are two common forms of the equipment. One consists of a relatively heavy cart of mass m_0 which is provided with small, light wheels that ride along horizontal steel rails.[1] The wheels are mounted in bearings to keep friction forces to a fairly low value. Attached to the front of the cart is a waxed-paper tape that passes over a cylindrical metal pulley P to a suspended weight hanger H. The total mass of the hanger and suspended weights is indicated in Fig. 6-1 by m.

Shown in dashed lines in Fig. 6-1 is a spark point S which is connected to a synchronous spark timer T. At periodic intervals, the spark timer produces a high voltage discharge which causes a spark impression to be made on the coated tape. The spark tapes are measured and analyzed in the same way as those ob-

[1] The Cenco track and car apparatus is of this type.

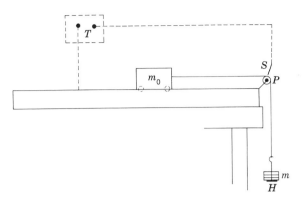

FIG. 6-1
Newton's second law
apparatus

tained in the free-fall experiment were. A number of spark traces may be made on a single 2-in.-wide waxed-paper tape. Each of these should correspond to a different value of suspended weight. All of the spark records produced in this experiment will be analyzed by a single computer program. The program will also compute the theoretical acceleration for each value of suspended weight. Graphs may then be made of both theoretical and experimental accelerations versus accelerating force.

The other form of Newton's second law apparatus makes use of a linear air track. This device is described in more detail in Appendix B. The air track consists of a hollow metal beam whose cross section has the shape of an inverted V. The surfaces of the V are very smooth. They contain many small holes through which air emerges under a positive pressure. The air produces an almost frictionless film on which a V-shaped glider moves. The glider takes the place of the cart of mass m_0 in Fig. 6-1. Attached to it is a light cord which passes over a low-friction pulley P, and is tied to a suspended mass m, as pictured in Fig. 6-1.

Aside from the replacement of the cart and rails by glider and air track, the major difference between the two forms of equipment is the manner in which the spark record is produced. In the air-track version of the experiment, the tape remains stationary on the metal air track while a metal spark point, which is attached to the glider, passes over it. The spark point is connected to a synchronous spark timer and produces spark dots along the tape at regular time intervals.

3. THEORY Figure 6-2 is a force diagram of the second-law apparatus illustrated in Fig. 6-1. It shows all the forces that act on the cart (or glider) and suspended weight. On the cart force diagram are the

FIG. 6-2
Force diagrams of cart (or
glider) and suspended
weight

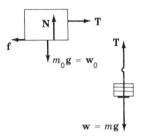

two vertical forces N and w_0, which must be equal and opposite to one another, since the cart is in equilibrium in the vertical direction; i.e., there is no acceleration in that direction. Directed horizontally are the force T, from the tension in the cord (or attached waxed-paper tape), and f, the friction force opposing the motion of the cart. In the horizontal direction we have from Eq. [6-4]

[6-5] $T - f = m_0 a$

The forces acting on the weight in the vertical direction are the tension force T and the weight force $w = mg$. The acceleration of the weight must be equal to the acceleration of the cart (in magnitude) since the two objects are connected by an inflexible cord (or tape). That this is so may be understood in the following way. When the cart moves to the right 1 cm, then the weight must drop down 1 cm also. If these displacements occur in 1 sec, then the velocities of both objects must be 1 cm/sec. If the velocity of the cart *changes* by 1 cm/sec, then likewise so must the velocity of the weight change by this amount. If these changes happen in 1 sec, then the average accelerations during that second of *both* bodies must be 1 cm/sec², etc.

Applying the second law, Eq. [6-4], to the vertical motion of the weight, we obtain

[6-6] $mg - T = ma$

Adding Eqs. [6-5] and [6-6] together, we have

[6-7] $mg - f = (m_0 + m)a$

The acceleration may be found by solving Eq. [6-7] explicitly for a,

[6-8] $a = \dfrac{mg - f}{m_0 + m}$

In the absence of friction, the acceleration is given by

[6-9] $$a = \frac{m}{m_0 + m} g$$

Equation [6-9] may also be written in terms of the weights, $w_0 = m_0 g$ and $w = mg$, by multiplying numerator and denominator by g

[6-10] $$a = \frac{w}{w_0 + w} g$$

The friction force f that appears in Eq. [6-8] may be determined approximately by trial and error. If a small value of suspended mass m is applied to the weight hanger, and the cart is given an initial push, it will be found either to accelerate or slow down. The suspended weight may then be changed accordingly. That value of suspended weight which causes no acceleration, i.e., for which the cart moves at constant speed, is f. This may be proved by considering Eq. [6-8] itself. When the acceleration a is zero, the numerator on the right-hand side of Eq. [6-8] must also be zero, that is, f must be equal to mg.

4. PROGRAMMING

There are many problems in mathematics, physics, and engineering in which it is necessary to manipulate arrays of numbers or variables. In this experiment, we are going to make use of the two-dimensional array

$$
\begin{array}{ccccc}
x_{11} & x_{12} & x_{13} & \cdots & x_{1m} \\
x_{21} & x_{22} & x_{23} & \cdots & x_{2m} \\
x_{31} & x_{32} & x_{33} & \cdots & x_{3m} \\
\cdots
\end{array}
$$

to store all of the information that we obtain from our spark-tape records.

Each index of the array x_{ij} in this experiment is used to label the spark trace data in a specific way. The first index, i, refers to the tape number; the tapes are numbered $1, 2, \ldots, n$. Thus, each *row* in the x_{ij} array corresponds to a single tape. The second index, j, which designates the *column* number in the array, also indicates the position of a spark dot on any of the tapes. For instance, the data in the *third* column represent measured positions of the *third* dots on all of the tapes. The x_{ij} array contains all measured displacements of spark dots relative to the $t = 0$ (first measured) dot on each tape.

In order to subject the spark trace data to the kind of analysis given the free-fall data in Experiment 4, it is necessary to retrieve and operate on the elements of the x_{ij} array one row at a time. These data must be processed mathematically by means of the procedure that was followed in Experiment 4 to find the average acceleration of the cart. While the experimental acceleration is being determined in this manner, the theoretical acceleration is to be computed with the aid of Eq. [6-8] or [6-9]. To accomplish these operations requires a knowledge of how two-dimensional arrays are treated in Fortran. This subject is discussed next.

Two-dimensional Subscripted Fortran Variables

Two-dimensional mathematical arrays, such as

$$
\begin{array}{cccccc}
a_{11} & a_{12} & a_{13} & \cdots & a_{1j} & \cdots \\
a_{21} & a_{22} & a_{23} & \cdots & a_{2j} & \cdots \\
a_{31} & \cdots & & & & \\
\vdots & & & & &
\end{array}
$$

are represented in Fortran by two-dimensional subscripted variables. The corresponding elements of a Fortran variable A are indicated in Fortran notation as A(I,J). The subscripts I and J are written in parentheses on the same line as A. They are separated by a comma. In similar fashion, the Fortran equivalent of a three-dimensional mathematical array a_{ijk} has elements denoted A(I,J,K). In Fortran II, only two- and three-dimensional arrays are permitted. In some versions of Fortran IV subscripted variables of seven and higher dimensions are allowed. We will confine our attention in this experiment to two-dimensional arrays and two-dimensional Fortran variables.

Before any two-dimensional subscripted variable appears in an executable statement in a Fortran program, the number of elements reserved in storage for it should have previously been specified in an appropriate DIMENSION statement. A DIMENSION statement that causes 10 rows and 50 columns to be reserved for the subscripted variable A is written[1]

DIMENSION A(10,50)

This nonexecutable specification statement must precede any statement that makes use of any element of A.

Two-dimensional Fortran variables are handled in exactly the same fashion as one-dimensional variables. However, because two indices are involved, it is necessary that definite integer values be

[1] The same result could be accomplished with a REAL statement, such as REAL A(10,50). The statement INTEGER A(10,50) would also reserve the proper amount of storage for A. However A would be treated as a fixed-point variable.

assigned to *both* indices when a two-dimensional variable is used in any executable statement. Otherwise Fortran will not know which element of the array to select from storage, and an error message will result. Thus, the arithmetic statement

C(1,2) = A(1,2) + B(1,2)

is a valid statement, whereas

C(I,2) = A(I,2) + B(I,2)

is valid *only if* a specific, positive integer value has previously been assigned to I in the program.

A simple program will illustrate the use of two-dimensional subscripted variables. In it, the element C(1,6) of one variable C is evaluated as the sum of the products of the elements in the first *row* of another variable A multiplied by the elements of the sixth column of third variable B. Multiplications of this kind are referred to as matrix multiplications, after the rules of the algebra of matrices. The program begins by reading in values of the required elements of A and B, and, after the required sum has been formed, concludes by printing out C(1,6). The program is:

```
DIMENSION A(6,6),B(6,6),C(6,6)
READ (5,8) (A(1,J), B(J,6), J=1,6)
C(1,6) = 0.0
J = 1
3 C(1,6) = C(1,6) + A(1,J)*B(J,6)
J = J + 1
IF (J − 6) 3, 3, 9
9 WRITE (6,8) C(1,6)
8 FORMAT (6F10.4)
```

Each of the three subscripted variables A, B, and C is defined as a 6 × 6 array in the DIMENSION statement. Not all of the elements in the array are used in the program. It is necessary to set C(1,6) equal to zero before the summation process is started. Otherwise, whatever value has been stored in the memory location which Fortran assigns to C(1,6) would be added to the sum. The complicated form of READ statement found in this program automatically performs the incrementing of index J from 1 to 6. This type of input-output statement is discussed in detail in the programming section of Experiment 8.

Figure 6-3 shows the flow chart of a program that may be used to perform an analysis of all the spark-tape data obtained in this experiment. As may be seen in the figure, subscripted variables

FIG. 6-3
Flow chart of a program
to analyze the spark tape
data produced by the ap-
paratus of Fig. 6-2

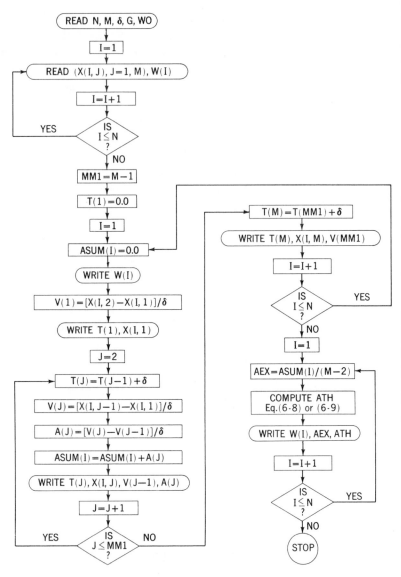

of two and more dimensions lead to fairly complicated flow charts
and to programs that comprise many statements. A great simpli-
fication in programming is afforded in such cases by the use of
nested DO loops. These are discussed in Experiment 9. However,
it is quite instructive to write a program of this type without DO
loops (*once!*) in order to understand the exact sequence of operations that must be performed.

The program of Fig. 6-3 first reads in N and M, the maximum
values of the subscripts I and J; δ,[1] the time interval between

[1] The value read in for δ should be assigned to a Fortran variable
named DELTA or DELT, etc.

measured spark impressions; G, the acceleration due to gravity; and WO, the mass of the cart (or glider). The letter W is again being used to indicate mass, as it was in Experiment 2. Next, the subscript I is set equal to 1. The program now initiates a counting loop which supervises the inputting of displacement data from spark-tape records, one trace at a time. The index I refers to the number of a particular tape trace. The data from the ith trace is read into the ith row of the x_{ij} array with the aid of the kind of READ statement that was employed in the matrix multi-plication program. The READ statement required in this loop is of the form

READ (5,100) (X(I,J), J=1,M), W(I)

It also causes the value of m_i to be read in and stored as W(I).

The main portion of the program consists of two loops, one *nested* inside the other. The outer loop increments I, the subscript designating the tape number. The initial statement inside this loop clears (zeroes) a subscripted variable in I named ASUM. This variable will be used to collect the sum of acceleration values computed from the displacement data on each tape. Later in the program, each element of ASUM will provide a value of the experimental acceleration for the corresponding tape.

The inner loop increments the subscript J, as the average veloc-ity and average acceleration during each interval are calculated. It also increments the time by δ. The velocities and accelerations are computed by taking differences as indicated in the following schematic diagram:

$t_1 = 0$	x_1		
		v_1	
$t_2 = \delta$	x_2		a_2
		v_2	
$t_3 = 2\delta$	x_3		a_3
		v_3	
$t_4 = 3\delta$	x_4	$\dots$	a_4
$\dots$	$\dots$		$\dots$

The subscripts in this diagram correspond to values of the index J. The sum of the a_j are also accumulated in this inner loop and stored as ASUM. Finally, the computed values of t_j, x_j, v_{j-1}, and a_j are printed out.

In the last program segment, the average experimental acceler-ation is calculated from the previously computed value of ASUM for each spark trace and assigned to AEX. The theoretical acceleration is computed with either Eq. [6-8] or [6-9] and is stored as ATH. The experimental and theoretical accelerations are then printed out for each trace along with the corresponding value of suspended mass, W(I).

5. PROCEDURE   **a**   Level the track. Attach the cord or tape to the cart (or glider). Pass the tape over the pulley and fasten it to the weight holder. Add sufficient mass to the weight holder so that the total suspended mass is approximately one-tenth the mass of the cart. The length of tape (or cord) should be adjusted so that the weight just touches the floor when the cart reaches the end of the track.

b   Connect the spark point to the synchronous spark timer. Adjust the position of the point so that it is near one edge of the waxed-paper tape and passes a seemingly continuous spark through the tape without any spurious arcing discharges to the track. Turn on the spark discharge and release the suspended weight. Turn off the discharge when the weight strikes the floor. Record the value of the suspended mass on the tape and assign it a number (1 for the first tape, etc.).

c   Decrease the suspended mass and repeat parts *a* and *b*, after displacing the spark point so that it produces a second trace parallel to and laterally shifted from the first one. Decrease the suspended mass and repeat again. Continue in this manner until a total of 6 to 10 traces has been made.

d   Lay each tape on a flat surface. Select a convenient, distinct spark dot from the early spark impressions. This dot is the origin. Place a meter stick on top of the tape with its markings touching the tape (i.e., lay it on its thin side with its markings down). Do not shift the meter stick while making measurements. Measure and record the position of every 6th or 12th dot (depending on whether the spark timer interval is $\frac{1}{60}$ or $\frac{1}{120}$ sec). Also record the tape number and the value of the suspended mass.

e   Referring to the flow chart of Fig. 6-3, write a Fortran program to analyze all the spark-trace data and compute experimental and theoretical values of acceleration. Use the measurements of part *d* as data in the program.

f   Plot on the same axes both experimental and theoretical accelerations versus suspended mass. As a variation of the experiment, keep the suspended mass fixed and change the mass of the cart by adding weights to it. Modify the program to compute the correct theoretical acceleration consistent with this data. Plot experimental and theoretical accelerations versus cart mass.

1. INTRODUCTION   Our study of rotational motion in this experiment will be limited to a consideration of the rotation of a rigid body about a fixed axis. A rigid body is one in which the distances between all particles are maintained constant during the motion of the body. The equation of motion that applies to this type of rotation is

[7-1] $\Sigma_\tau{}^{\text{net}} = I_0 \alpha$

This equation is the analog of Newton's second law. In Eq. [7-1], I_0 is the moment of inertia of the body about the rotation axis, τ^{ext} is the external torque about this axis, and α is the angular acceleration of the body.

The moment of inertia of a body about an axis is defined by the equation

[7-2] $I_0 = \sum_{i=1}^{n} m_i r_i{}^2$

In this definition, the body is assumed to consist of n point masses. In Eq. [7-2], r_i is the perpendicular distance from the ith point mass, m_i, to the axis of rotation. For continuous distributions of mass, the summation is replaced by an integration

[7-3] $I_0 = \int r^2 \, dm$

Here, r is the perpendicular distance between the axis and an infinitesimal element of mass, dm. The two cases represented by Eqs. [7-2] and [7-3] are illustrated in Fig. 7-1. The rotation axis in both diagrams is perpendicular to the plane of the page.

In a number of special cases, involving highly symmetric rigid bodies, simple expressions may be derived for the moment of inertia about an axis of symmetry. For a homogeneous, solid cylinder or disk, for example, the moment of inertia about its symmetry axis is

[7-4] $I_{\text{cyl}} = I_{\text{disk}} = \frac{1}{2} m r^2$

FIG. 7-1
System of point-mass particles (a); and continuous distribution of mass (b)

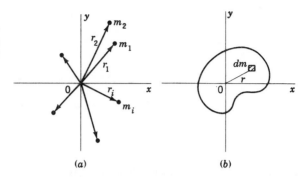

(a)   (b)

In Eq. [7-4], r is the radius of the cylinder (or disk). The length of the cylinder does not appear in the formula. Equation [7-4] is derived in the general physics references cited in the bibliography. The derivation involves a simple application of Eq. [7-3].

2. APPARATUS

The essential features of the rotation wheel apparatus required in this experiment are shown in Fig. 7-2. It consists of a metal flywheel which rotates on low-friction ball bearings about a horizontal axle. The wheel is composed of two sections: (1) a uniform disk of large radius (r_2) and (2) a cylindrical hub of smaller radius (r_1). The hub is equipped with a small protruding pin P on which a cord may be anchored. The cord is wound several times around the hub and attached to a suspended weight hanger H.

A dimensioned force diagram of the apparatus is presented in Fig. 7-3. The diagram shows the radii r_1 and r_2 and the initial height h of the suspended weight above the floor. The weight force mg and the tension in the cord T are also shown. Note that the same tension force T acts both on the weight (upwards) and on the wheel (downwards).

FIG. 7-2
Rotation wheel apparatus

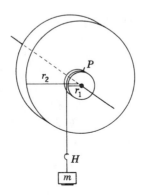

FIG. 7-3
Force diagram of the rotation wheel apparatus

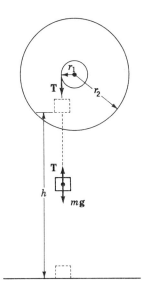

Weights, whose total mass is m, are suspended from the cord
(should the mass of the weight hanger be included?). The weights
are released with zero velocity at time $t = 0$ as a stopwatch is
started. The time t_i required for the weights to fall the distance
h is determined for each of a number of applied load masses.
An experimental value of the angular acceleration α is calculated
for each m from the values of h, r_1, and t, as discussed in the next
section. Each of these values of α is to be compared with an α
calculated theoretically from the system parameters r_1, r_2, m_1, m_2,
and m. The theoretical value of α is improved by adding a correction for the frictional torque at the axle.

3. THEORY   An experimental value of α may be deduced from the measured
values of h, r_1, and t as follows. The angle in radians through which
the wheel turns is related to r_1 and h by

[7-5] $\theta = \dfrac{h}{r_1}$

This relation is valid since h represents both the distance through
which the weights fall and the distance traveled by a point on
the circumference of the hub (assuming no slipping of the cord).
Using the rotational analog of the constant-acceleration formula
[4-4],

[7-6] $\theta(t) = \omega_0 + \tfrac{1}{2}\alpha t^2$

89

with $\omega_0 = 0$ (no initial angular speed), we obtain for α

[7-7] $\alpha = \dfrac{2\theta}{t^2}$

Applying Eq. [7-5] to eliminate θ, we have

[7-8] $\alpha = \dfrac{2h}{r_1 t^2}$

The theoretical angular acceleration may be derived starting from Eq. [7-1]:

[7-9] $\tau^{\text{ext}} = r_1 T - r_1 f = I_0 \alpha$

There is a frictional retarding torque included in Eq. [7-9]. It is equal to the radius r_1 multiplied by a force f, which would be needed to overcome the frictional torque when applied at the hub of the wheel. Experimentally, f is measured by suspending weights from a cord wrapped around the hub. That particular value of weight which keeps the wheel turning at constant angular speed, once started, is f.

The equation of motion of the weight is provided by Newton's second law. If the positive y axis is taken to point downward, the equation may be written

[7-10] $\Sigma F_y = mg - T = ma$

The tangential acceleration of a point on the rim of the hub is related to the angular acceleration of the wheel by

[7-11] $a = r_1 \alpha$

This tangential acceleration of the hub is the same as the downward acceleration of the weight since the cord is assumed not to slip. It is anchored by the pin P to the hub. Combining Eqs. [7-10] and [7-11], we have

[7-12] $mg - T = mr_1 \alpha$

Solving Eq. [7-12] for T and using the result to eliminate T in [7-9], we obtain

[7-13] $\alpha = \dfrac{mg - f}{mr_1 + I_0/r_1}$

The moment of inertia of the wheel may be computed by means of Eqs. [7-2] and [7-4]. The moment of inertia of the hub is given by

[7-14] $I_{\text{hub}} = \frac{1}{2}m_1 r_1{}^2$

and of the large-radius flywheel, by

[7-15] $I_{\text{wheel}} = \frac{1}{2}m_2 r_2{}^2$

Since the two cylinders share a common axis of rotation, the moment of inertia of the composite flywheel by Eq. [7-2] must be

[7-16] $I_0 = \frac{1}{2}(m_1 r_1{}^2 + m_2 r_2{}^2)$

This is the value of I_0 that should be substituted into Eq. [7-13] in the computation of angular acceleration.

4. PROGRAMMING   Input and output operations on card readers and line printers are most commonly brought about in Fortran by READ, WRITE, and PRINT statements. Each of these is accompanied by an appropriate FORMAT statement. The FORMAT instructs the computer how to read the data found on the punched cards or in what form to present the computed output. We proceed now to discuss these input-output and FORMAT statements in detail.

Fortran   Immediately following the words READ and WRITE in Fortran
Input-output   IV there are two numbers separated by a comma and enclosed
Statements   by parentheses. The first of these numbers refers to the particular input-output device which is to be used. Frequently employed designations for card reader and line printer are 5 and 6, respectively. The second of the numbers inside the parentheses identifies the associated FORMAT statement. In Fortran II, only the FORMAT statement number is specified, and no parentheses are used. Typical READ statements in the two versions of Fortran which accomplish the same input of data are

Fortran II: READ 10, J, A, B, C
Fortran IV: READ (5,10) J, A, B, C

Here, three floating-point variables, A, B, and C, and one integer variable, J, are to be read. They could all be punched on a single card. For example, let us assume that we want to locate the fixed-point number representing the value of J in the first two-column *field* (we refer to a set of adjacent columns that correspond to a single datum entry as a *field*) and three floating-point numbers which represent values of A, B, and C in the following

three 10-column fields. A FORMAT statement which will accomplish the correct reading of a data card prepared in this
manner is

10 FORMAT (I2,F8.3,F8.3,F8.3)

As discussed in Experiment 2, the I2 identifies a two-column
integer, or fixed-point, field. The F specification refers to a floating-point number without an exponent (power of 10). A more
compact notation for the above FORMAT statement would be

10 FORMAT (I2,3F8.3)

The number 8 to the left of the decimal point in each of the
F8.3 specifications designates an eight-column field. The number
3 to its right specifies that three digits should be placed at the
right of the decimal point in the number stored in memory. When
reading in data under an F8.3 format, the 3 would ordinarily not
be important, since the actual location of the decimal point in
any number punched in the eight-column field takes precedence.
If, however, no decimal point is punched in the field, Fortran then
assigns the decimal according to the F8.3 FORMAT specification.
For example, suppose that we have punched the number 12345678
in columns 3 through 10 of one data card which is read under
FORMAT statement 10 above. It is interpreted as the number
12345.678 and it is stored as such in the memory area alloted
to A. If, instead, the number 123456.7 were punched in these
columns, then that would be the value of A stored in memory.

In some circumstances, it might be more convenient to punch
a single data entry on each of a set of data cards. This is facilitated
by the character "/" inside the FORMAT statement. It indicates
the beginning of a new *record*, i.e., a new data card. In the previous
example, if numbers corresponding to the values of J, A, B, and
C were punched in the leftmost columns of *separate* data cards
in the same size and type fields previously described, they could
be read under a FORMAT specification

10 FORMAT (I2/F8.3/F8.3/F8.3)

Again, there is a more compact way to write the preceding
FORMAT statement. If the last specifications within a FORMAT
are enclosed by parentheses, then data continue to be read under
these format codes after all specifications in the statement have
once been used. Thus, an appropriate FORMAT statement which
most concisely accomplishes this input is

10 FORMAT (I2/(F8.3))

The I2 causes the reading of an integer number punched in the leftmost two columns of the first data card. The slash mark (/) requires the reading of a new card. The last specification, F8.3, is enclosed by parentheses, meaning that all successive records read under this FORMAT statement will be treated as though they contained a single eight-character floating-point number in columns 1 to 8. Any blank spaces in the first eight columns of any of these records (cards) are interpreted as zeroes. We could have called for the reading of *two* 8-column floating-point numbers on all cards after the first by specifying

10 FORMAT (I2/(2F8.3))

The E format code calls for the reading of a data entry which contains a floating-point number multiplied by a power of 10. The power of 10 exponent is usually indicated on the data card by the letter E followed by a signed number. Examples of numbers which might be read under an E format code are

$-1.23456E+12$

and

$-123456E7$

Notice that in the second of these numbers we have contracted the $E+07$ exponent notation to a simpler E7 form. Also allowed are the contractions $-123456+07$ and $-123456+7$. Any of these forms could be read under an E12.3 specification provided that it was located within the 12 columns indicated by this format code. The 12 to the left of the decimal point in E12.3 indicates a 12-column field. The 3 to its right specifies the number of places to the right of the decimal point in the number put in storage. Once again, as in the case of the F format code, the 3 is overridden in an actual data entry that includes a decimal point. The second number above read in under an E12.3 specification would be interpreted as the number -123.456×10^7, since it does not contain a decimal point.

The rules for output in Fortran are very similar to those required for input. The WRITE statement has the same form as the READ statement. It also must be associated with a FORMAT statement whose number is indicated by the second entry in the parentheses (following the word WRITE). Suppose for example that we have stored in memory the following values for the variables J, A, B, and C: 12, 1.23456, 12.3456, and -123.456. The pair of output statements

```
     WRITE (6,10) J, A, B, C
10 FORMAT (I3, 3F8.3)
```

would produce the following line of printed output

12bbb1.235bb12.346 − 123.456

where b stands for a blank space. We are assuming here that the computer being used rounds off those portions of the output data that are not printed. The I3 specification rather than an I2 was necessary because the first character of an output record is ordinarily not printed. As explained below, it is used to control the printer carriage. Output records are shifted one space to the left, and the first character is dropped before printing.

The usual output statement in Fortran II is the PRINT statement. It is similar in form to the Fortran II READ statement. The PRINT statement that corresponds to the WRITE in the example cited in the preceding paragraph is

```
PRINT 10, J, A, B, C
```

PRINT is also a valid output statement in Fortran IV.

Literal and X Specifications

Two more valuable output specifications are the literal, or Hollerith, and column-skipping format codes, indicated by H and X, respectively.[1] The number of characters to be printed or the number of columns to be skipped are indicated by an integer immediately preceding the H or X. The H and X specifications may be included among I, F, and E format codes in a FORMAT statement. They are used in this fashion to label and space data entries.

The Hollerith format code causes the specified number of keyboard characters to be printed out *exactly* as they are written in that many columns following the letter H. As a simple example, let us consider the following pair of statements:

```
     WRITE (5,10) I, R(I)
10 FORMAT (8H TRIAL =, I3,4X,6HREVS. =, F7.0)
```

If the values of I and R(I) happen to be 2 and 4312.3 when these statements are executed, they will produce the line of output

TRIAL =   2   REVS. =   4312.

A Hollerith specification often stands by itself in a FORMAT statement to supply headings for output listing, etc. Apostrophes

---

[1] The X specification is also used on input to cause a desired number of columns on the data card to be ignored by the computer.

enclosing the desired literal message may be used in place of the H format code in Fortran IV. The following example illustrates the use of H and X specifications. Let us assume that we have previously computed values for centripetal force as a function of speed according to Eq. [5-3] and have stored these values in one-dimensional arrays named FCENT and V.

```
    WRITE (6,100)
    WRITE (6,101) (V(I), FCENT(I), I = 1,N)
100 FORMAT (1H1,14H FORCE(DYNES) ,10X,'SPEED(CM/SEC)'/)
101 FORMAT (1H ,E14.6,10X,E14.6)
```

The first pair of WRITE and FORMAT statements cause two 14-character data headings, separated by 10 spaces, to be printed. The second pair of WRITE and FORMAT statements require the printing of N pairs of V and FCENT data floating-point numbers with exponents in data columns 14 characters wide, the columns being separated by 10 spaces. The 1H1 specification in statement 100 is for the control of the printer carriage. When the first character in an output record is 1, 0, blank, or +, the computer instructs the line printer to print according to the conventions

1   Begin printing on a new page
0   Skip a line and begin printing
b   Print on the next line
+   Print on the same line

It is always advisable to use a carriage control character as the first format specification in an output record to ensure correct printer line spacing. The control character itself is not printed.

At the right-hand end of FORMAT statement 100 we have written a slash mark (/). Its function is to cause a line to be skipped between the printed column headings and the output data. The line skipping comes about in the following way. The slash causes a new record to be started. Therefore, the printer moves down one line. The next output record starts at the left-hand end of still another line. Thus, a one-line spacing between the heading and the first data entry results.

The second WRITE statement in the above example has a fairly complicated form. This type of I/0 statement, which will be discussed at greater length in Experiment 8, is very useful for effecting the input and output of subscripted variables. When it is executed, this statement first sets I equal to 1, and then causes V(1) and FCENT(1) to be printed on a single line according to

FORMAT statement 101. It increments I by one, and repeats the printing of pairs of V(I), FCENT(I) until V(N), FCENT(N) have been printed. The READ statement in the matrix multiplication program previously discussed in Experiment 6 is another example of the use of the implied DO. Notice that the value of N which is required as the limit of the variable I is supplied in the same READ statement. In that READ statement, the computer is first instructed to read in a value for the integer variable N, and then to read in values of the subscripted floating-point variable X(I) for $1 \leq I \leq N$.

FIG. 7-4
Flow chart of a program to compute angular accelerations and print out results

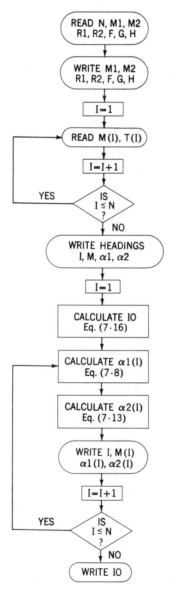

The program of Fig. 7-4 performs the analysis of data obtained with the rotation wheel apparatus. The only new, noteworthy feature of the program is its use of a wide variety of format specifications, such as the Hollerith specification (which provides column headings and labels), along with the E and X format codes discussed in this section, and the F and I specifications, which have been described previously. A sample program is presented in Appendix D as Fig. D-5.

5. PROCEDURE   **a**   Measure and record the diameters of the flywheel and hub. Their masses will be found either stamped on the cylinders or written on labels attached to the apparatus. Record these values.

b   Determine the force f in the following fashion. Tie a loop in one end of a cord and anchor it on the pin P. Wrap the cord several times around the hub and attach its free end to a light weight hanger. Add just enough weight to the suspended weight hanger so that, once started, it drops at *constant* speed. The total suspended weight, including that of the weight hanger, is then equal to f. This is so since f must provide just enough torque about the axle to counteract the frictional torque and keep the wheel turning at constant angular speed.

c   Attach a weight hanger whose mass is less than 100 g to one end of a cord. The cord should be long enough to extend from the pin P, when it is in a horizontal orientation, to the floor. Tie a loop in the other end of the cord, secure it to pin P, and wrap the cord carefully without overlap around the hub. Add enough mass to the weight hanger to bring the total suspended mass to 100 g.

d   Steady the suspended weight and release it at the instant a stopwatch is started. Determine the precise time interval required for the weight to drop to the floor. Repeat the determination for suspended masses of 200, 400, and 600 g. Measure and record h.

e   Referring to Fig. 7-4, write a Fortran program that will analyze the data and compute experimental and theoretical values of the angular acceleration for each of the suspended weights. The experimental α should be computed by means of Eq. [7-8]. The moment of inertia should be computed with the aid of Eq. [7-16] and used in Eq. [7-13] to compute the theoretical angular acceleration for each value of load mass, m. The program should cause all data and results, both experimental and theoretical, to be printed out in labeled, neatly tabulated form. Use as many different FORMAT specifications as you can in your program to accomplish this. Note that in the flow chart of Fig. 7-4 experimental and theoretical angular accelerations are indicated by α_1 and α_2, respectively.

f   One form of the rotation wheel apparatus is equipped with a spark timing mechanism similar to the one used in Experiment 4.[1] A waxed-paper disk is affixed to the flat surface of the wheel. A circular spark trace is produced on the disk as the wheel turns. If this type of rotation wheel apparatus is being used, the data obtained with it may be conveniently analyzed by a program similar to the one written in Experiment 4. Write the program and analyze your data with it.

[1] The Cenco rotational inertia apparatus is of this type.

EXPERIMENT 8
YOUNG'S MODULUS

1. INTRODUCTION All materials, even such strong ones as structural steel, when subjected to sufficient force, experience a measurable deformation. The subject of elasticity deals with the behavior of materials under the action of various forces and torques. In this experiment, we will study the deformation of slender rods and thin wires acted upon by tensile (stretching) forces.

To make quantitative our discussion of elasticity, we define *stress* as force per unit area, and *strain* as the relative deformation that it produces. There are different types of stress and strain. As mentioned above, we are interested in the essentially one-dimensional case of tensile stress-strain. Tensile stress is defined as the stretching force per unit cross-sectional area. A wire in tension is shown in Fig. 8-1. The tensile force is labeled **T**. The cross-sectional area is labeled A; the initial length of wire, y_0; and the elongation of the wire, Δy. In terms of these quantities, the strain is $\Delta y/y_0$. The stress is T/A.

For most structural materials, the strain produced is linearly related to stress for small to moderate values of stress. This may be seen in the typical stress-strain graph presented in Fig. 8-2. The graph is representative of the elastic behavior of metals and alloys. From point O, the origin, to point A, called the *proportional*

FIG. 8-1
Deformation of a wire under the action of a tensile force

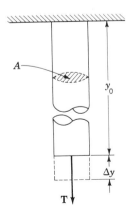

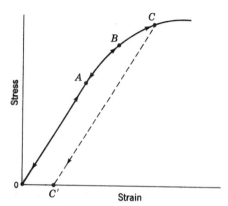

FIG. 8-2
Stress-strain graph of a
typical metal or alloy

limit, the graph is linear. At point A, the curve begins to deviate from linearity. Between points A and B, the curve is reversible and single-valued; i.e., as the stress is reduced, the original curve is retraced back to point O. When the stress has been entirely removed, the wire returns to its initial length y_0. Point B is called the *elastic limit*. Beyond point B, this is no longer the case. The curve followed by the material as the stress is diminished, CC' in the diagram, is different from the curve it follows while the stress is increased, $OABC$. When the stress has been reduced to zero, there remains a permanent deformation of the wire. In Fig. 8-2, this permanent deformation is equal to the strain OC'. The behavior of the material in the region of the curve beyond point B is referred to as *plastic* deformation.

In the linear region of Fig. 8-2, where stress is proportional to strain, we may define a constant of proportionality M_y by

[8-1] $$M_y = \frac{\text{stress}}{\text{strain}} = \frac{F/A}{\Delta y/y_0} = \frac{Fy_0}{A\,\Delta y}$$

M_y is known as *Young's modulus*. It is characteristic of the material of which the wire is composed. Young's modulus is a basic elastic quantity which appears in many equations in the mechanics of solids.

2. APPARATUS

The apparatus needed to measure M_y is shown in Fig. 8-3. In the diagram, a thin wire is shown suspended from the top of a rigid support frame. It passes through a chuck which is tightened about it. At its lower end, the wire is attached to a weight hanger. The reason for the chuck and the mirror platform which rests on it is to facilitate the measurement of Δy. The mirror, platform, telescope, and scale are the components of a device called an *optical*

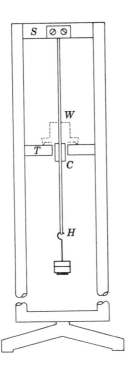

FIG. 8-3
Young's modulus appara-
tus. Components are: sup-
port frame (S), wire (W),
chuck (C), mirror-platform
support table (T), and
weight hanger (H). The
outlines of the mirror are
seen behind the chuck.

lever. The optical lever magnifies the elongation of the wire (optically) by a factor which is ordinarily in the range of about 20
to 500. This is necessary because even when the wire is subjected
to values of stress near the proportional limit, its deformation is
seldom greater than about 1 mm. To measure directly distortions
of a wire of this order of magnitude would require a very sensitive
measuring microscope. Even with a good microscope, the percent
accuracy of Δy measured directly is generally rather poor. Therefore, we employ the light beam of the optical lever to magnify
the change in length of the wire. Details of its operation are
discussed in the next section. The commercially produced Cenco
and Sargent-Welch Young's modulus instruments are both of
this type.

3. THEORY   The computation of M_y is performed using Eq. [8-1]. As previously
discussed, the difficult factor to measure is Δy. Details of the
optical lever used to facilitate its measurement are shown in Fig.
8-4. A small mirror M is located at the right-hand end of the
metal platform P. On the bottom of P are three small pointed
metal feet. The single rear foot of P rests on the chuck C, which
is fastened to the wire. The two front feet rest in a groove in a
support table T which is clamped to the support frame.

FIG. 8-4
Details of the optical lever:
mirror (M), metal platform
(P), support table (T), viewing telescope (V), and
scale (OR)

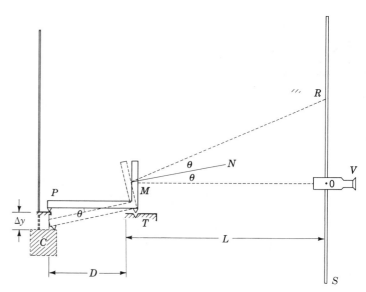

As the wire is stretched by the load weights, the chuck on which P rests drops by Δy. This causes the platform to tilt through an angle θ, as indicated by the dashed lines in Fig. 8-4. The mirror also rotates through θ. Let us assume that initially the platform is in a horizontal orientation and that the mirror surface is in a vertical plane. Light from the zero position on the vertical scale S strikes the mirror perpendicularly. It is reflected back upon itself to the viewing telescope V. When the wire is pulled down a distance Δy, the mirror rotates through an angle θ. The light beam incident on the telescope now originates from a point higher on the scale. This point is labeled R in Fig. 8-4. Henceforth, we will refer to the distance OR simply as R (the scale *reading*).

The law of reflection from a smooth surface requires that the angle made by the incident ray (RM) with the mirror normal (MN) be equal to the angle between the reflected ray (MO) and the normal. Therefore, the angle between incident and reflected rays (RMO) is equal to 2θ.

In the derivation that follows, the small-angle approximation

[8-2] θ (radians) $= \sin \theta = \tan \theta$

will be required. It is valid within about $\frac{1}{2}$ percent for angles smaller than $10°$. Applying Eq. [8-2] to triangle RMO, we have

[8-3] $2\theta = \tan 2\theta = \dfrac{R}{L}$ or $\theta = \dfrac{\frac{1}{2}R}{L}$

Applying Eq. [8-2] again, this time to the small triangle formed by the two positions of the platform (solid and dashed lines in

Fig. 8-4), we obtain another relation involving θ,

[8-4] $\theta = \tan \theta = \dfrac{\Delta y}{D}$

Combining the two expressions for θ, Eqs. [8-3] and [8-4], and solving for Δy, we have

[8-5] $\Delta y = \left(\dfrac{\frac{1}{2}D}{L} \right) R$

As a check on the validity of the small-angle approximation which has been used in the derivation of Eqs. [8-3], [8-4], and [8-5], let us consider a specific set of experimental data that might be taken with a typical optical lever system. Typical values of the lever parameters D and L are 5.0 and 200.0 cm. According to Eq. [8-5], a 1.0-mm elongation of the wire would cause a scale reading change of 8.0 cm in this case. Equations [8-3] and [8-4] both yield a value of 0.02 radians for the angle θ that corresponds to a Δy of 1.0 mm. For an angle θ of this magnitude, the small-angle approximation of Eq. [8-2] is certainly justified. This may be verified by referring to a set of tables of sines, tangents, and degrees to radians.[1]

The computational problem that arises in the analysis of our optical lever data is similar to the one encountered in Experiment 4. We have here a set of successive readings of the telescope scale for equal increments of added load mass. If we treated these data by simply summing the differences between successive measurements and dividing by the number of increments times the load increment, we would again be effectively discarding all readings but the first and last. In order to avoid this difficulty, we again apply the method of differences. It is used here to compute the average value of elongation per unit of added load mass.

The average scale deflection change per increment of applied load mass is calculated first with the assistance of the method of differences. This result is multiplied by the optical lever ratio of Eq. [8-5], i.e., by $\frac{1}{2}D/L$, to obtain the elongation of the wire per increment of added load mass. When this value is divided by g, the elongation per dyne of applied *weight* is obtained.[2] This is the

---

[1] See for example, "Saunders Short Tables," W. B. Saunders Company, Philadelphia, 1967, or any edition of the "Handbook of Chemistry and Physics," Chemical Rubber Publishing Company, Cleveland.

[2] Assuming that cgs units are being used. If mks units are employed, the ratio obtained will be equal to the average elongation in meters per newton of added load weight.

ratio $\Delta y/F$ that appears in the formula for Young's modulus, Eq. [8-1]. It must be multiplied by the cross-sectional area of the wire and then divided into the initial length y_0 to obtain M_y. If ΔR is used to indicate the average change in scale reading per gram of added load (calculated by the method of differences), then, according to the preceding discussion, M_y must be related to ΔR by

$$[8\text{-}6] \quad M_y = \frac{y_0}{A(\tfrac{1}{2}D/L)(\Delta R/g)} = \frac{2y_0 g L}{AD\,\Delta R}$$

4. PROGRAMMING The flow chart of a program to compute Young's modulus using the method of differences to compute the ratio of scale deflection change per unit of applied load is given in Fig. 8-5. It is very similar to the program written in Experiment 5 to compute the centripetal force acting on a rotating bob. That program also makes use of the method of differences. A new feature which appears in the program associated with this experiment is the DO loop. The program itself is very straightforward, and does not

FIG. 8-5
Flow chart of a program to compute Young's modulus using the method of differences

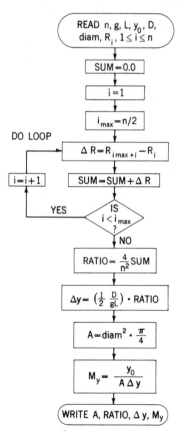

introduce any other new concepts. The remainder of this section will be devoted to a discussion of some of the properties of the DO statement.

The DO Statement   One exceedingly important and useful feature of Fortran is the DO statement or DO loop. With it one can accomplish in a single statement all the steps that were previously required to initialize, increment, and test a counter or index so that a portion of a program could be repeated a desired number of times. The simplest form of the DO statement is

DO [Statement number] Index = M1,M2

The statement number referred to above is the number of some statement below the DO in the program. The *index* is some integer variable. M1 and M2, which may either be fixed-point numbers or variables, specify the *initial* and *final* values assumed by the index in the DO loop. The use of the DO is illustrated below.

For a specific example, let us consider the simple DO statement

DO 7 I = 1,25

Notice that all numbers and variables which appear in a DO statement must be in fixed-point form. The integer 7 refers to the number of a statement which may be found somewhere in the program below the DO. The DO statement requires that the index I first be set equal to 1, and that all statements below the DO statement itself up to and including statement 7 be executed in order. This group of statements is referred to as the *range* of the DO.

After all statements in the range of the DO have been executed once, the DO causes the index I to be incremented by 1 (i.e., I is set equal to 2). The same set of statements is then executed again with the new value of I. This is repeated for all integral values of I in the interval between 1 and 25. After the 25th pass through these program steps, control is transferred to the statement immediately following statement 7.

A convenient feature of the DO is its ability to skip numbers as it increments its index. This is called for by specifying a third integer or integer variable after the equals sign in the DO statement. It is set off by a comma from the preceding two integers. This integer indicates the increment that is to be added to the index. If it is omitted, as in the example above, the increment is taken to be 1. To illustrate the use of this kind of DO statement, let us assume that we want to perform a group of program statements down to statement 7 for all *odd-integral* values of an index I between 1 and 25. The DO statement with which we may accomplish this is

DO 7 I = 1,25,2

We may replace any of the integer numbers which appear in a DO statement with an integer variable, provided that this variable has been assigned a value in a previous program statement. There is one exception, and that is the statement number which follows the DO, e.g., 7 in the example above. It must always be written as an integer, which refers to a numbered statement located somewhere in the program after the DO statement. Thus, we could write

DO 7 I = INITL, LAST

It is necessary that values of the integer variables INITL and LAST have been read in, defined, or calculated in earlier program statements. All statements between the DO and statement 7 will be executed in order for each value of I from INITL to LAST.

There are a number of rules regarding the kind of statements that may be used in various positions in the DO loop. The most important of these are the requirement that the first statement following the DO itself be executable, the prohibition on any statement causing a transfer of control as the last statement in the range of the DO, and the regulations concerning valid transfers of control into and out of the DO loop. All of these rules will be discussed at greater length in the next experiment.

As a simple example of the use of the DO statement, let us consider a program segment that inputs values from data cards into the 2 one-dimensional subscripted arrays named X and Y. Suppose that we have 25 pairs of (X,Y) data to read and that one pair of data is to be found on each of 25 cards in 10-column-wide data fields. A program of the type that we have previously employed to accomplish the desired input is

```
    DIMENSION X(25), Y(25)
    I = 1
  7 READ (5,100) X(I), Y(I)
    I = I + 1
    IF (I − 25) 7, 7, 9
  9 ...
100 FORMAT (2F10.0)
```

The use of a DO statement simplifies the programming of the loop considerably. With the DO, the program would be written

```
DIMENSION X(25), Y(25)
   DO 7 I = 1, 25
 7 READ (5,100) X(I), Y(I)
100 FORMAT (2F10.0)
```

There is an even simpler, more highly condensed way of inputting this data by means of a single Fortran statement (apart from the necessary DIMENSION and FORMAT statements). This type of input (or output) statement permits the incrementing of subscripts to be performed by the statement itself. Such an input (or output) statement is, for obvious reasons, often referred to as an *implied* DO statement. It is especially useful when data are to be read into (or printed out from) only a portion of an array.[1] For the problem at hand, this kind of READ statement could be written

```
READ (5,100) (X(I), Y(I), I = 1, 25)
```

Combined with the DIMENSION and FORMAT statements of the above examples, it would achieve the desired data input.

If we wanted to read the X data first, followed by the Y data, we could accomplish this with the single implied DO statement

```
READ (5,100) (X(I), I = 1, 25), (Y(I), I = 1, 25)
```

If the FORMAT statement numbered 100 in the previous examples accompanied this READ, the X and Y data would be read from data cards with two 10-column floating-point entries per card. It is possible, of course, to read just one array with a single implied DO statement. For example, if we were interested only in the Y array, we could cause it to be read with the statement

```
READ (5,100) (Y(I), I = 1, 25)
```

5. PROCEDURE   **a**   Attach a thin steel wire to the top of the support frame. Pass the wire through the chuck and tighten the jaws of the chuck around it. Fasten the lower end of the wire to a suspended weight hanger.

b   Place on the hanger a tare weight of sufficient magnitude to straighten the wire and remove any slight kinks it might have (if there

[1] This use of the implied DO statement is discussed in R. M. Lee, "A Short Course in Fortran IV Programming," p. 84, McGraw-Hill Book Company, New York, 1967. The implied DO is also treated in detail in D. D. McCracken, "A Guide to Fortran IV Programming," p. 83. John Wiley & Sons, Inc., New York, 1965.

are severe kinks in the portion of the wire between the support frame and the chuck, that length of wire should be replaced by one without these defects). For a very thin wire (on the order of 0.1 mm in radius), several hundred grams will be sufficient. For thicker wires, one or more kilograms of tare weights may be necessary. Excessive tare weight will deform the wire beyond its elastic limit, so do not apply a larger tare than necessary. Note that the value of the tare weight will not be used in the calculations. We are interested only in the deflection produced (i.e., the *change* in length of the wire) per *increment of added applied weight.*

c   Measure the diameter of the wire carefully with a micrometer caliper. Record it. Measure and record the length of wire between support frame and chuck. This is the value of y_0 that will be used in the calculation of Young's modulus.

d   Set the viewing telescope at a distance of about 1.5 m from the mirror. It should be at approximately the same height as the mirror. Measure and record the distances D and L of Fig. 8-4.

e   Focus the telescope eyepiece on the cross hairs. Point the tele-scope toward the mirror. Adjust the length of its eyepiece-to-objective distance to bring into focus the image of the scale which is reflected from the mirror back to the telescope. This adjustment is rather delicate and may require a bit of patience on the part of the experi-menter. It will probably require that the orientation of the mirror and of the telescope be shifted several times to bring the scale into view.

f   Once the scale image has been located and brought into focus in the telescope, adjust the inclination of the mirror so that the lower portion of the scale is seen at the cross hairs. Then, as the applied load is increased, much of the remainder of the scale will be seen.

g   Record the initial scale reading. Add a load increment of 200 g (for a thin wire, 1 kg for a thick wire). Record the new scale reading. Repeat this procedure until an even number of readings have been obtained. The method of differences requires that an even number of readings (including the initial one) be made.

h   Write a Fortran program to compute Young's modulus from these experimental data. Refer to the flow chart of Fig. 8-5. The program should read in and print out all required experimental data. It should then compute the average scale deflection per increment of added load (in grams) by the method of differences explained in Experiment 5. This ratio should be converted to the elongation of the wire per dyne of applied load weight, as previously discussed in Section 3. Finally, the program should compute Young's modulus with the aid of Eq. [8-6] and print out all results in a neatly labeled and tabulated form.

i   The experimental value of Young's modulus should be compared with its handbook value. A simple error analysis following the procedure outlined in Experiment 2 should be performed. Do so by estimating the error in each factor of Eq. [8-6]. Your final reported answer should include a plus or minus estimate determined from the error analysis.

EXPERIMENT 9
THE BALLISTIC PENDULUM

1. INTRODUCTION   The ballistic pendulum is a device used to measure the velocity of a projectile. After an inelastic collision, the projectile lodges in the pendulum and causes it to rise a distance which is determined by the initial projectile velocity. In the analysis of the experimental data to calculate this velocity, the concepts of conservation of energy and momentum are required. Once the initial velocity of the projectile has been determined, its range and time of flight in a projectile motion experiment may be computed theoretically (using the equations of rectilinear motion for each angle of projection). This experiment, therefore, is a valuable one for illustrating many basic principles of mechanics.

The essential features of the ballistic pendulum are shown in Fig. 9-1. A projectile of mass m is fired with an initial velocity v_0 toward a heavy block or carriage, whose mass M is much larger than m. The projectile lodges itself in the carriage, and the two move off immediately after collision with a velocity V. The principle of conservation of momentum requires that the momentum in the system before collision be equal to the momentum immediately afterwards

[9-1] $$mv_0 = (M + m)V$$

The carriage and projectile swing upward and come to a stop instantaneously when the center of gravity of the system has risen

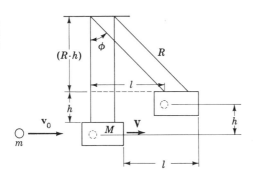

FIG. 9-1
Elementary ballistic pendulum

to a height h above its initial position. The distance h is related to V, m, and M by the energy conservation principle. The initial kinetic energy of the carriage and projectile must be equated to their gain in potential energy when they come to a stop,

[9-2] $\quad \frac{1}{2}(M + m)V^2 = (M + m)gh$

When we solve Eq. [9-2] explicitly for V, we obtain an equation which is independent of mass,

[9-3] $\quad V = \sqrt{2gh}$

Combining Eqs. [9-1] and [9-3], eliminating V, and solving for v_0 in terms of m, M, g, and h, we have

[9-4] $\quad v_0 = \dfrac{M + m}{m} \sqrt{2gh}$

All of the quantities on the right-hand side of Eq. [9-4] can readily be measured in the laboratory. Equation [9-4] may be used to compute the projectile velocity v_0 once these measurements have been performed. This velocity is required in the computations of range and time of flight.

2. APPARATUS   The ballistic pendulum apparatus is illustrated in Fig. 9-2. In its simplest form, it consists of a spring gun S, a suspended carriage C, and a ball projectile B. In addition, the apparatus may be provided with a toothed-rack (R) and pawl (P) mechanism to catch the carriage at the high point of its swing.[1] It is convenient to have this mechanism, for then the height h may be determined statically by direct measurement.

FIG. 9-2
Experimental ballistic pendulum apparatus

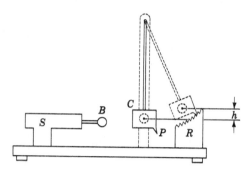

[1]A mechanism of this type is provided with the Cenco-Blackwood ballistic pendulum apparatus.

In some forms of the ballistic pendulum apparatus, there is no rack and pawl mechanism to hold the pendulum at the high point of its swing. In such cases, either the angle ϕ or horizontal displacement l shown in Fig. 9-1 may be measured, along with the length R of the suspending wire. If the angle ϕ is measured, l may be computed from it by means of the trigonometric relation (see Fig. 9-1)

$$l = R \sin \phi$$

Once l has been found, the distance h may be calculated with the aid of the pythagorean theorem. Applied to the triangle in Fig. 9-1 that contains sides of length $(R - h)$ and l, and hypotenuse R, it may be expressed

$$(R - h)^2 = R^2 - l^2$$

This equation may be solved to find h.

The projectile motion portion of this experiment is a variation of the usual study made with the ballistic pendulum apparatus.[1] The carriage is swung out of the way, the front end of the apparatus is blocked up to make an angle θ with horizontal, and the projectile is fired toward the far corner of an obstacle whose upper surface is level with the projectile before it is fired. This is illustrated in Fig. 9-3. The height of the corner of the obstacle above the floor is labeled H. The horizontal distance from projectile to corner is labeled L. The range of the projectile for different elevation angles will both be measured and computed.

3. THEORY The initial projectile velocity is determined with the aid of Eq. [9-4], as previously discussed. The carriage is then removed from

FIG. 9-3
Ballistic pendulum apparatus arranged for the study of projectile motion

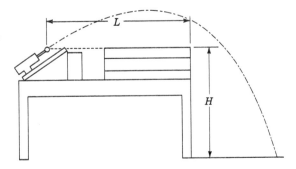

[1] This experiment was suggested by prob. 4-18 of "Physics," D. Halliday and R. Resnick, John Wiley & Sons, Inc., New York, 1966.

the flight path of the projectile, and the spring gun is inclined at an angle θ with horizontal. Depending on the magnitudes of the elevation angle θ and the initial projectile velocity v_0, the projectile when fired may land either on top of the obstacle or on the floor. In either case, the usual equations of uniformly accelerated motion

[9-5] $v(t) = v_{0_y} - gt$

and

[9-6] $y(t) = v_{0_y}t - \frac{1}{2}gt^2$

may be employed to describe its motion in the y direction.

Let us first consider the case in which the projectile lands on the obstacle. We will use a standard system of coordinates (y axis upward and x axis to the right) with the origin at the initial position of the projectile. Resolving the initial velocity into its x and y components in the usual fashion, we have

[9-7] $v_{0_x} = v_0 \cos \theta$

and

[9-8] $v_{0_y} = v_0 \sin \theta$

When the projectile lands on the obstacle, its y coordinate is again equal to zero. Substituting Eq. [9-8] into Eq. [9-6], we have at that instant

[9-9] $0 = (v_0 \sin \theta)t - \frac{1}{2}gt^2$

Solving Eq. [9-9] for t, and neglecting the solution $t = 0$, we obtain for the time of flight

[9-10] $t = \dfrac{2v_0}{g} \sin \theta$

We have disregarded the solution $t = 0$ of Eq. [9-9]. We might consider briefly what its significance really is; i.e., is it actually a valid solution of Eq. [9-9] which has some physical interpretation? The answer clearly is *yes*. It corresponds to the initial point on the trajectory of the projectile, when the projectile has not yet departed from the origin. This solution gives us no additional useful information concerning the subsequent motion of the projectile, and we are justified in neglecting it.

There is no acceleration component in the x direction. Therefore, to calculate the range of the projectile, we need only multiply the time of flight, just computed in Eq. [9-10], by the x component of velocity, computed in Eq. [9-7]. The result is

[9-11] $$x = \frac{2v_0^2}{g} \sin \theta \cos \theta = \frac{v_0^2}{g} \sin 2\theta$$

This equation is often referred to as the horizontal range formula. It applies only to the horizontal range of a particle when it has returned to its initial level.

When the projectile returns to the ground at a level different than that from which it was fired, the equations for time of flight and range are slightly more complicated. The time of flight is calculated by solving the quadratic equation

[9-12] $$-H = (v_0 \sin \theta)t - \tfrac{1}{2}gt^2$$

This equation results when we substitute $y = -H$ and $v_{0y} = v_0 \sin \theta$ in Eq. [9-6]. The solution for t is

[9-13] $$t = \frac{v_0 \sin \theta}{g} + \left[\left(\frac{v_0 \sin \theta}{g} \right)^2 + \frac{2H}{g} \right]^{1/2}$$

The range is again found by multiplying this time by the constant x component of velocity, $v_0 \cos \theta$, as in the previous case.

4. PROGRAMMING

In this experiment, we are going to write a program for the range of the projectile of the ballistic pendulum illustrated in Fig. 9-3 as a function of its angle of elevation. The program consists of two separate DO loops. The first loop computes the range according to Eq. [9-11], starting with an elevation angle of zero, and increments the angle coarsely in 5° jumps. When the range computed in this fashion becomes greater than L, control is transferred to a second DO loop. It performs a finer incrementing of the elevation angle, in 1° steps.

After returning to the last elevation angle in the first loop for which the projectile still lands atop the obstacle, the second loop begins its computations. It calculates time of flight and range for both cases: (a) the projectile landing on the obstacle, and (b) landing on the floor beyond. To be able to write a program of this kind, we must know how control is transferred to and from DO loops, and how their indices behave in such transfers. We proceed now to discuss these matters.

The DO statement introduced in Experiment 8 is subject to a number of strict rules, especially those concerned with transfer of control into and out of the loop. We are going to consider the restrictions on such transfer of control, both with regard to a single DO loop and to a set of *nested* DO loops.

Transfer of Control to and from a DO Loop

A DO loop sets up and increments a counter, called the *index* of the DO loop, each time the loop is executed. It is important that no statement in the range of the DO disturb this counter, so that a confusing value of the index of the DO results. Nor may the variables which define the initial and final values of the counter be modified *in any way whatsoever* in any statement in the range of the DO.

For example, in the statement

DO 8 I = INIT, LAST

the variables INIT and LAST specify the initial and final values the index I is to assume. We could use either of these variables in an arithmetic statement such as

K = LAST − I + 1

This statement does not *change* the value of LAST, although it does make use of its value. Forbidden is a statement such as

LAST = LAST − 1

which *modifies* the variable LAST. The same prohibition exists on a statement which would modify the counter (i.e., the index I).

The first statement after the DO itself must be an executable statement. Specifically prohibited are such statements as DI-MENSION, FORMAT, and CONTINUE (described below). However, a second DO *is* permitted in this position. When several DO statements are stacked in this manner, they are referred to as a *nest* of DO's. Nested DO loop configurations will be discussed later.

The last statement in the range of a DO (the one whose number appears immediately after the word DO) may not involve a transfer of control, nor may another DO statement be located in this position. Neither the GO TO nor the IF may be written as the last statement in a DO loop. In order to avoid this restriction and to make the final *operation* performed in the DO loop a transfer of control, its last statement is made a CONTINUE statement.

The CONTINUE statement consists of just that one word. It is usually preceded by the statement number specified in some DO statement. It is a neutral (or *dummy*) statement that provides a proper termination for the DO loop, and permits the preceding statement in the loop to be of any kind, including GO TO and

IF statements. The CONTINUE statement is often used as the last statement in a DO loop whether or not the next to last statement involves a transfer of control. However, if the CONTINUE statement is not really necessary, its automatic use as the last statement in the range of a DO loop is not a recommended programming practice.

Since the DO statement must increment and test its index each time the loop is executed, the means of transfer of control *into* the loop must be carefully regulated. In general, a DO loop may only be entered through the DO statement itself. Control may be transferred out of the loop in any statement except the last. When control passes from the loop *before* it has been satisfied, the index retains whatever value it has at the time of transfer. The loop is said to be *satisfied* when it has been executed the required number of times. In the previous example, that would be the case after the loop was executed with its index I equal to the value of the variable LAST. Finally, the index is cleared. It must be redefined before it is next used. The following example illustrates valid transfer of control to and from a DO loop

```
    GO TO 15
15 DO 17 I = 1, N
    IF (X(I).GT.L) GO TO 18
17 CONTINUE
18 . . .
```

Transfer into the DO loop is accomplished by the GO TO statement, which passes control to the DO statement itself. The IF statement sandwiched between statements 15 and 17 causes control to be transferred outside of the DO loop to statement 18, if the value of the subscripted variable X(I) is greater than the value of the variable L.

The Logical IF Statement Another feature of Fortran IV which we have not as yet discussed appears in this program segment. It is the *logical* IF statement. It is not available with Fortran II compilers. The general form of the logical IF statement is[1]

IF(X.Operator.Y) Statement

The operators in the logical IF statement are set off by periods from the variables and/or constants upon which they operate. The

[1] The logical IF statement is discussed in R. W. Southworth and S. L. DeLeeuw, "Digital Computation and Numerical Methods," p. 62, McGraw-Hill Book Company, New York, 1965.

usual set of operators, known as *relational* operators, is given below. When the expression inside the parentheses is satisfied for the current values of X and Y, the statement written in the logical IF is executed.

Operator	Significance
LT	Less than
LE	Less than or equal
EQ	Equal
NE	Not equal
GT	Greater than
GE	Greater than or equal

The statement following the parentheses may not be another logical IF, nor may it be a DO. It must be an *executable* statement. The logical IF in the above program is equivalent to the three-address IF

IF(X(I) − L) 17, 17, 18

Nested DO Loops   It is permissible to enclose one DO loop inside another DO loop. This is especially convenient when working with two-dimensional arrays and matrices. When one DO loop is enclosed within another, the loops are referred to as a *nest* of DO loops (or as nested DO loops). It is important that the enclosed (inner) DO loop be *entirely* enclosed, that is, have all of its statements inside the enclosing (outer) DO loop. This is necessary so that the indices of both loops are properly incremented as each completes a pass through its range. Otherwise, one DO would try to increment its index and return to the beginning of its range before the other had completed all of its statements, as in the example

```
    DO 10 I = 1, 20
    DO 15 J = 1, 20
10 READ (5,100) V(I,J)
15 WRITE (6,101) V(I,J)
```

First, I would be set equal to 1 and control would be transferred to the second DO. It would set J = 1 and try to perform all statements down to 15. However, before that could happen, statement 10 would have to be executed. This statement is the end of the range of the first DO, which is a signal for control to be transferred back to that DO and the index I to be incremented.

Therefore, the second DO could not be completed without interference from the first, and the situation would be very confused. This nesting configuration is forbidden.

An example of a pair of properly nested DO loops is found in the following program segment:

```
    DO 20 I = 1, 4
    WRITE (6,100) I
    DO 10 J = 1, 5
    READ (5,101) V(I,J)
 10 WRITE (6,102) J, V(I,J)
 20 CONTINUE
100 FORMAT (1HO, 'ROW', I2)
101 FORMAT (F10.0)
102 FORMAT (1H , I2, F12.4)
```

When executed, the above program proceeds in the following manner. First, I is set equal to 1, and the literal

ROW 1

is printed on a single line, after the printer has skipped one line (due to the 1HO carriage control character). The data corresponding to $V(1,1), V(1,2), \ldots, V(1,5)$ are read from data cards, one entry per card. Immediately after each card is read, its value is printed out (or stored for later printing) on another line along with a number that corresponds to its column in the V array. The column number, J, is printed to the left of the value of the $V(1,J)$. After the five elements of row 1 of the V array have been read and printed in this manner, the index I is incremented by 1, one row is skipped by the printer, and the literal

ROW 2

is printed. Five more cards are read and the values of $V(2,J)$ they contain are printed as previously described. The whole process is repeated for rows 3 and 4 of the array.

It is possible to have nested DO loops end on the same statement. In the preceding program both loops could have terminated on the same WRITE statement. The same results would be produced by the program (using identical FORMAT statements)

```
    DO 10 I = 1, 4
    WRITE (6,100) I
    DO 10 J = 1, 5
    READ (5,101) V(I,J)
 10 WRITE (6,102) J, V(I,J)
100 . . .
```

FIG. 9-4
Nested DO loop configura-
tions: (a) forbidden nesting;
(b) allowed nest with two
separate terminal state-
ments; (c) allowed nest
with inner and outer DO's
ending on the same
statement

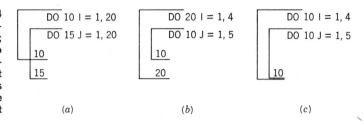

(a)   (b)   (c)

The three program segments discussed in this and the preceding several paragraphs each contain a pair of DO loops. Two of these programs show allowed nests of DO statements, and one, a prohibited DO nesting. Schematic representations of these nested DO-loop configurations are given in Fig. 9-4.

FIG. 9-5
Flow chart of a program to calculate the theoretical range of a projectile fired from various elevation angles. The projectile may strike an obstacle or the floor beyond it.

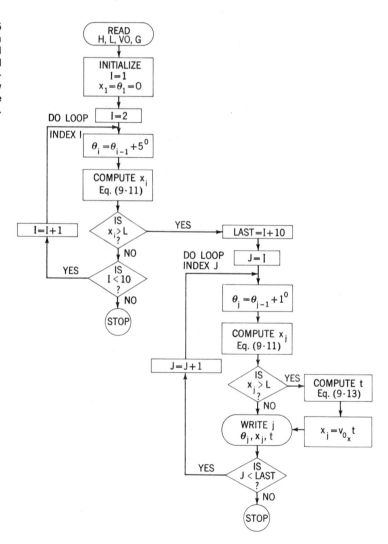

The flow chart of a program which computes the horizontal range of a projectile for various elevation angles is presented in Fig. 9-5. The program involves an allowed transfer of control out of one DO loop and into another. It makes use of the value of the index of the first loop at the time of transfer in setting the values of the initial and final limits of the index of the second DO loop.

The index of the first loop, I, is incremented by 1 each time the loop is executed. The angle of elevation, θ, is also incremented in the program each time the loop is executed. It is increased by $5°$ in each pass. The range[1] of the projectile is computed by means of Eq. [9-11]. A comparison is made of the computed range with the distance L. If the range is greater than L, control is transferred out of the first DO loop. If the computed range for all angles of elevation from $0°$ through $45°$ remains smaller than L, the program is terminated by a STOP statement.

When control is transferred out of the first DO loop, the value of its index, I, is used as the initial value of the index, J, of the second loop. The final value of the index J is the value of the variable LAST. LAST is equated to the value of I plus 10.

In the second DO loop, the angle θ is initially set equal to the *next to last* value computed for it in the first loop (the last value of θ for which the projectile landed on the obstacle) plus $1°$. Each time the loop is executed, the angle θ is incremented by $1°$. The range of the projectile is then computed by means of Eq. [9-11]. Next, a comparison of the range with L is made. If the range is less than L, control is passed to the final statement of the DO. The index J is incremented and the loop is again executed. If the computed range is greater than L, control is passed to an arithmetic statement which computes the time of flight by means of Eq. [9-13]. The range is then computed by multiplying this time by the x component of the projectile velocity. In this manner, ten values of the horizontal range of the projectile, no matter whether it lands on the obstacle or on the floor, are computed by the loop for successive $1°$ increments of the elevation angle.

5. PROCEDURE

a Level the ballistic pendulum apparatus. Weigh and record the masses of the projectile and carriage. The mass of the carriage may be found stamped on its surface.

b Determine the position of the center of gravity of the loaded carriage with the projectile inside it. This may be done by removing

[1] The projectile *range* and the *range* of a DO should not be confused.

the pendulum from the apparatus and balancing it on a fulcrum. The center of gravity is located directly above the fulcrum when the pendulum is balanced on it. This determination may already have been performed and the position of the center of gravity indicated by a small pointer located on the carriage.

c   Measure the height of the center of gravity of the pendulum while it is at rest in its vertical orientation. Fire the projectile into the carriage and measure the height of its center of gravity at the high point of its swing. Repeat this determination several times and average the measurements. Subtract from this average height the height of the center of gravity of the pendulum at rest to find h. Compute the initial velocity of the projectile by means of Eq. [9-4]. This calculation may also be built into the computer program associated with the flow chart of Fig. 9-5, if desired.

d   Set up the ballistic pendulum apparatus as indicated in Fig. 9-3. Place a cardboard or wooden box or some other rectangular obstacle topped by a thin, soft cushion or mat on the table opposite the spring gun, as shown in the figure. Adjust the height of the obstacle so that the top of the cushion or mat is at the same level as the projectile before it is fired.

e   Block up the apparatus so that it is inclined at an angle of 5° with horizontal. Adjust the height of the top of the obstacle so that it is initially level with the projectile. Put a sheet of $8\frac{1}{2} \times 11$-in. paper on top of the mat. Lay a piece of carbon paper face down on it. Increase the angle of elevation in 5° increments until the projectile lands atop the obstacle.

f   Fire the projectile and note where it lands. Adjust the position of the paper so that the projectile lands near its center. Fasten it to the mat. Fire the projectile several times. Remove the carbon paper and find the average position of the projectile impressions on the $8\frac{1}{2} \times 11$-in. paper. Measure the distance from the projectile (before firing) to this average position.

g   Increase the elevation of the spring gun by 5° and repeat part f. Continue in this fashion until the projectile passes over the rectangular obstacle without striking it. If the initial velocity of the projectile is insufficient to propel it over the obstacle at any elevation angle between 0 and 45°, terminate the procedure. Move the spring gun toward the obstacle. Determine whether the projectile has sufficient velocity to pass over the obstacle by firing it from an elevation angle of 45°. As Eq. [9-11] shows, this is the angle that produces the maximum horizontal range. (Verify this statement.) Once a satisfactory value of L has been determined in this way, perform the 5° incrementing of elevation angle described above.

h Now aim at the far corner of the obstacle and experimentally determine the minimum angle θ for which the projectile just clears the corner. Because of small variations in the initial projectile velocity from run to run, there will be a spread in the values obtained for θ. The correct θ will have been found when about half of the shots fired pass over the obstacle and half land on it. With the gun elevated at this angle θ, determine the average horizontal range of the projectile both when it lands on the mat and when it lands on the floor. Measure L and H. Record all these values.

i Refer to the flow chart of Fig. 9-5. Write a Fortran program to compute the range of the projectile for 5° increments in the elevation angle while it lands on the mat and then for 1° increments, as previously discussed in Section 4. Compare the experimental results obtained in parts *f*, *g*, and *h* with these computed values.

j As a simple variation of the programming procedure, write a Fortran program that avoids the transfer between DO loops. It should merely increment θ by 1° or 2° and perform the range computations as indicated in the second DO loop of Fig. 9-5 (the one with index J). Compare the experimental range measurements with these computed values.

THE SIMPLE PENDULUM— DIFFERENTIAL EQUATIONS

1. INTRODUCTION   Newton's second law of motion in its most elementary vector form (applied to a body or particle of mass m) is

[10-1] $\mathbf{F}^{net} = \Sigma \mathbf{F} = m\mathbf{a}$

Equation [10-2] may also be expressed as three component equations

[10-2] $\Sigma F_x = ma_x$
$\Sigma F_y = ma_y$
$\Sigma F_y = ma_z$

It is the fundamental law of classical mechanics and serves as the starting point for the derivation of most of the relations of classical dynamics.

In most cases of interest the net force acting on a particle or body may be written as a function of time and position

[10-3] $\mathbf{F} = \mathbf{F}(\mathbf{r}, t) = \mathbf{F}(x, y, z, t)$

In Eq. [10-3], the position of the body is specified at any given instant of time by its three Cartesian coordinates x, y, z or by its position vector $\mathbf{r}$. This situation is illustrated in Fig. 10-1.

For the special case of one-dimensional motion in the x direction, Eq. [10-3] becomes

[10-4] $F_x = F_x(x, y, z, t)$

The acceleration of the body is, by definition, the second time derivative of position, $d^2\mathbf{r}/dt^2(d^2x/dt^2$, in one dimension). Therefore, in all of the above cases, Newton's second law produces one component differential equation for each Cartesian coordinate, e.g.,

[10-5] $m\dfrac{d^2x}{dt^2} = F_x(x, y, z, t)$

This equation follows from Eqs. [10-2], [10-4], and the definition of acceleration as the second time derivative of position.

125

FIG. 10-1
Components of position
and force acting on a
particle of mass m. The
position of the particle is
indicated by the vector r,
whose components are
x, y, and z.

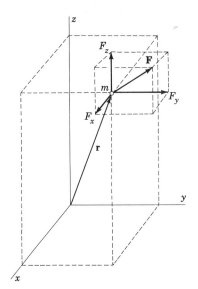

Equation [10-5], under special circumstances, reduces to a much simpler form. In the case of one-dimensional motion in the x direction, for instance, when there is no y and z dependence, the equation may be written

[10-6] $$m\frac{d^2x}{dt^2} = F(x,t)$$

Furthermore, there are a number of situations of interest in which the force on the right-hand side of Eq. [10-6] has no explicit time dependence. This is true of an important class of forces which depend only on the coordinates of the particle called *conservative* forces. Conservative forces, such as the force of gravity and the elastic restoring force exerted by a stretched or compressed spring, are discussed in greater detail in the general physics references in the bibliography.

In the case of one-dimensional forces of this type, Eq. [10-6] may be expressed as

[10-7] $$m\frac{d^2x}{dt^2} = F(x)$$

We will be concerned in this experiment only with simple differential equations of the form of Eq. [10-7].

Two important, elementary differential equations are produced when either of the conservative forces

[10-8] $$F(x) = -mg$$

or

[10-9] $$F(x) = -kx$$

FIG. 10-2
Gravitational force acting
on a particle of mass m
near the surface of the
earth

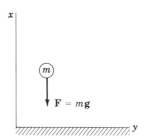

is substituted for $F(x)$ in Eq. [10-7]. The first of these is the gravitational force which acts on a particle of mass m near the surface of the earth. It is illustrated in Fig. 10-2. In the figure, the x axis is directed upward. The force points in the negative x direction, and is of constant magnitude. It produces a constant downward acceleration of the particle.

The force given in Eq. [10-9] is the elastic restoring force exerted by a stretched or compressed *ideal* spring on an object which is attached to it. The constant of proportionality k in Eq. [10-9] is called the *force constant* of the spring. In Fig. 10-3, a bob of mass m is shown fastened to the end of the spring. Once displaced, the bob oscillates back and forth continuously. This vibration, which has a sinusoidal time dependence, is known as *simple harmonic motion*. It is a consequence of the differential equation of motion, which we will now study.

The differential equations that result when the forces of Eqs. [10-8] and [10-9] are substituted into Eq. [10-7] are

$$[10\text{-}10] \quad m\frac{d^2x}{dt^2} = -mg \quad \text{or} \quad \frac{d^2x}{dt^2} = -g$$

and

$$[10\text{-}11] \quad m\frac{d^2x}{dt^2} = -kx \quad \text{or} \quad \frac{d^2x}{dt^2} = -\frac{k}{m}x$$

Both of these differential equations have important applications in engineering and physics. They occur in a variety of problems in electricity and mechanics. Although exceedingly simple, they serve as first approximations to many more complicated real

FIG. 10-3
Elastic restoring force
acting on a simple har-
monic oscillator of mass
m attached to an ideal
spring of force constant k

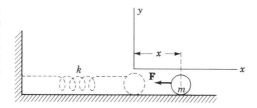

problems. Because of their simplicity and ease of handling mathematically, they are often used as the starting point for more involved mathematical treatments.

Both Eqs. [10-10] and [10-11] possess mathematical *solutions* which are familiar, elementary functions. By the term solution to a differential equation we mean a function which satisfies the equation, i.e., which may be correctly substituted into it. For instance, Eq. [10-11] has a solution of the form

[10-12] $x_1(t) = \sin\left(\sqrt{\dfrac{k}{m}}\,t\right)$

That this is a solution to the differential equation should be verified by substitution of Eq. [10-12] into Eq. [10-11]. However, Eq. [10-12] is not the *only* solution to Eq. [10-11]. Another solution is

[10-13] $x_2(t) = \cos\left(\sqrt{\dfrac{k}{m}}\,t\right)$

This also should be verified. Other solution functions also exist.

Equations [10-10] and [10-11] are examples of *linear* differential equations. A linear differential equation is one of the form

[10-14] $0 = a_n(t)\dfrac{d^n x}{dt^n} + a_{n-1}(t)\dfrac{d^{n-1}x}{dt^{n-1}} + \cdots + a_0(t)x(t) + b(t)$

The *order* of a differential equation is the order of the highest derivative that appears in the equation. As written, Eq. [10-14] is an nth-order linear differential equation. An important property of linear differential equations of order 2 and higher is that, when $b(t)$ is zero, *all* possible linear combinations of solutions are also solutions. This means in the case of Eq. [10-11] that not only are x_1 and x_2 solutions, but also $3x_1 + 4x_2$, $x_1 - x_2$ etc. In general, there will be an infinite number of solutions to this kind of equation. Choose some particular linear combination of x_1 and x_2 and show that it satisfies Eq. [10-11]. Its general solution is

[10-15] $x(t) = cx_1(t) + dx_2(t)$

Since Eqs. [10-10] and [10-11] both come from physical problems, there should be a *unique* solution for the position of the particle represented by each differential equation at each instant of time, i.e., in mathematical terms, each equation should possess *one* correct solution function $x(t)$. This is, in fact, the case. However, in order to choose the correct solution from among the infinite number of possible solution functions, we must specify more information. This information ordinarily takes the form of initial condition equations on the position and velocity of the particle.

The theory of differential equations requires that, in order to determine uniquely the solution of a second-order equation, values of both the function x and its first derivative dx/dt must be specified at some instant of time. Physically, this is equivalent to stating the position and velocity of the particle at that time. The usual procedure is to specify the position and velocity at time $t = 0$. The corresponding equations for $x(0)$ and $v(0)$ are referred to as *initial condition* equations, or, simply, as initial conditions.

Up to this point we have discussed the origins of differential equations in various physical problems, solution functions for these equations, and, finally, initial condition equations which are necessary to single out the unique solution from all of the possible solution functions. Rather straightforward methods of solution are available for handling linear differential equations. Frequently, however, these are quite complicated, and it must not be assumed that all *simple-appearing* differential equations have simple solutions. Still more involved are nonlinear differential equations, i.e., those that cannot be written in the form of Eq. [10-14]. The problem that we are going to investigate in this experiment is a case in point. Its physical basis is the equation of motion of a simple pendulum. The innocent-looking differential equation that must be solved in this problem is

[10-16] $$\frac{d^2\theta}{dt^2} = -\frac{g}{l}\sin\theta$$

Even as simple a differential equation as [10-16] cannot be solved in terms of elementary functions. That is why computers are so widely used to solve differential equations. In this case, the solution requires either elliptic integrals or some numerical approximation technique. The latter approach is the one that we will take here. In Section 3 we will describe a numerical point-by-point method of solution. The corresponding program will be discussed in Section 4.

2. APPARATUS

The apparatus needed to study the motion of the simple pendulum consists of the pendulum itself, a stroboscope light source, and a camera whose shutter permits time exposures of the order of 1 sec. The experimental setup is illustrated in Fig. 10-4. It is essential to have a dark, nonreflecting backdrop behind the appa-

FIG. 10-4
Apparatus to photograph
the motion of a simple
pendulum: stroboscope
(S), camera (C), and
pendulum (P)

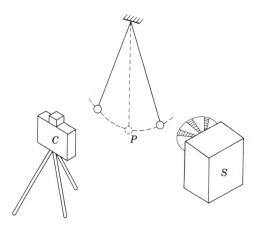

ratus to obtain clear photographs with good contrast between the lighted bob and the background.

The stroboscope is set to flash at some standard frequency in the range of 20 to 50 flashes/sec. The pendulum bob is released from a given angle with vertical, and multiple exposures are made as it moves through one-half of a vibration, i.e., from a maximum angle on one side of the vertical equilibrium position to an equal angle on the other side. The pendulum is stopped, and an additional exposure is made of the bob in its equilibrium orientation for reference in analyzing the photograph. The developed and printed photograph (Polaroid film is very convenient to use in this experiment and considerably shortens the required laboratory time) is secured to a flat surface. For each flash exposure of the pendulum, the angle between it and the vertical is carefully measured. These experimental θ versus t data are to be plotted superimposed on the theoretical θ versus t graph. The latter graph is plotted with the aid of numerical results obtained from a computer solution of Eq. [10-16]. The details of the required program are discussed in Section 4.

Before the experimental data can be graphed, the point on the multiflash photograph corresponding to $t = 0$ must be determined. The most satisfactory way to accomplish this is to assign $t = 0$ to the vertical equilibrium position. Linear interpolation may be used to determine the time at which the exposure nearest this equilibrium position occurred. Once this has been done, all other times are determined. This is so because the frequency of the stroboscope is assumed to be known accurately, and we need only add or subtract an integral number of stroboscope periods to compute the time at each point.

To illustrate the method of calculating the time corresponding to each exposure, we have made a line drawing that represents

FIG. 10-5
Typical (hypothetical)
multiflash photograph of a
vibrating simple pendulum
(drawn from data pro-
duced by the computer
program of Fig. 10-7)

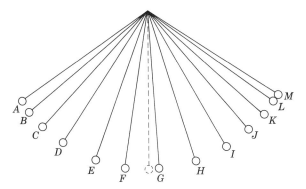

a hypothetical multiflash photograph. The numerical data used
in constructing the drawing, Fig. 10-5, were obtained from a
computer solution of Eq. [10-16]. The individual exposures are
indicated by letters A through M in the figure; the vertical equilibrium position is shown with dashed lines.

The pendulum starts its swing at the left, and swings toward
the right. The angle between it and the vertical in each orientation
is given in Table 10-1. Nearest the equilibrium position are exposures F and G. The corresponding angles are $-9.3°$ and $+3.9°$.
Their difference is $13.2°$. If we denote the stroboscope period by
T, and if we choose $t = 0$ at the equilibrium position, then, by
linear interpolation, exposure G must have occurred at
$(3.9/13.2)T$. For example, if the stroboscope frequency happened
to be exactly 8.00 flashes/sec, T would be 0.125, and exposure
G would have taken place at $t = 0.037$ sec. Exposure H would

TABLE 10-1
Experimental
θ–versus–t Data

Exposure	Angle, deg	Time, sec
A	−55.8	−0.713
B	−51.2	−0.588
C	−43.8	−0.463
D	−34.4	−0.338
E	−22.9	−0.213
F	−9.3	−0.088
G	3.9	0.037
H	17.6	0.162
I	30.4	0.287
J	40.6	0.412
K	48.7	0.537
L	54.4	0.662
M	57.3	0.787

have occurred 0.125 sec later, i.e., at $t = 0.162$ sec etc. The times indicated in the right-hand column of Table 10-1 were all determined in this manner.

3. THEORY The differential equation of motion of the simple pendulum follows from the rotational analog of Newton's second law

[10-17] $\tau^{\text{net}} = I_0 \alpha$

This equation is identical to Eq. [7-1]. In it, I_0 is the moment of inertia about the rotation axis, τ^{ext} is the net torque of all external forces, and α is the angular acceleration. The moment of inertia of a simple pendulum, regarded as a point mass tied to a cord of length l, is

[10-18] $I_0 = ml^2$

Referring to Fig. 10-6, we see that when the instantaneous displacement of the pendulum is $+\theta$, there is a torque about the rotation axis given by

[10-19] $\tau = -mgl \sin \theta$

The torque appearing in Eq. [10-19] is called a *restoring* torque, since for positive θ's (counterclockwise displacement angles) the torque is negative (it tends to produce clockwise rotation), and vice versa. When this torque is substituted into Eq. [10-17], and both sides are divided by ml^2, the result is the differential equation previously written as Eq. [10-16], i.e.,

[10-20] $\alpha \equiv \ddot{\theta} = -\dfrac{g}{l} \sin \theta$

We have employed the definition of angular acceleration as the second time derivative of angular displacement θ in obtaining this result (time derivatives are indicated by dots placed above the quantities to be differentiated).

There are many numerical procedures available for solving differential equations. An especially simple solution method is possible for equations of the form of Eq. [10-7]. It consists of point-by-point extension of the solution from an initial point at which the function x and its first derivative are known. These quantities are specified by the initial conditions associated with the problem. The solution is extended to an adjacent point by calculating the position, velocity, and acceleration at the new point from the values of these quantities at the original point.

132

This procedure is repeated for each of a sequence of equally spaced points at which the solution is desired.

In order to keep our discussion of the solution procedure on a practical level, let us consider the simple pendulum again. Its motion is described by the differential equation

[10-21] $a_x = \ddot{x} = -\dfrac{g}{l}\sin x$

In this equation, we have used $x(t)$ to represent $\theta(t)$, the angle between the pendulum and vertical. Equation [10-21] has the same form as Eq. [10-7]. Let us assume that the acceleration of gravity is exactly 9.8 m/sec^2. For ease of computation, we will take the length of the pendulum to be 0.980 m, i.e., 98.0 cm. Under these circumstances, its differential equation of motion is

[10-22] $\ddot{x} = -10.00 \sin x$

In order to have a simple set of initial conditions with which to work, we will assume that the pendulum is released with zero initial velocity from an initial angle of 57.296°, i.e., *one radian*. The corresponding initial conditions are

[10-23] $x(0) = 1.000$

and

[10-24] $\dot{x}(0) = 0.000$

The first decision that we must make prior to beginning our numerical solution is on the size of the interval δ between solution points. This is an important decision, and δ should be chosen with care. The smaller the value of δ, the smaller is the computational error which is introduced into the solution. At the same time, however, the number of computations and, therefore, the amount of computer time required, increase as δ is reduced. As usual in computations of this kind, when a fairly large number of repeated calculations are performed, a compromise between accuracy of result and amount of computer time needed must be made. For the sake of simplicity in the example that we are considering, we will use a rather coarse value of δ for the interval between solution points. To make the computation easy, let us take $\delta = 0.100$ sec. There are a great many elaborate methods which may be used to improve the accuracy of a numerical solution of a differential equation. However, these methods are beyond the scope of this book, and they will not be discussed further here. A number of different numerical solution procedures for differential equations

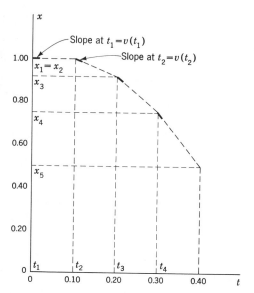

FIG. 10-6
Illustrating the numerical
solution of the simple
pendulum differential
equation: $\ddot{x} = -10 \sin x$
with initial conditions,
$x(0) = 1.000$, and
$v(0) = 0$

are described in detail in the numerical analysis references cited in the bibliography.

Figure 10-6 illustrates the solution procedure that will be employed to solve Eq. [10-22] subject to the initial conditions stated in Eqs. [10-23] and [10-24]. The region in which the solution is to be found is shown divided into equally spaced intervals 0.100 sec apart in the figure. The corresponding times are $t_1 = 0.000$, $t_2 = 0.100$, $t_3 = 0.200$, etc. (all times are in seconds).

The initial point on the solution function is at $t_1 = 0$, where $x = 1.000$, by Eq. [10-23]. At this point, according to Eq. [10-24], the slope must be equal to *zero*. This is indicated by the short, heavy bar drawn *horizontally* from the initial point. By continuing this bar to the point $t_2 = 0.100$, the solution is extended to the second point. In terms of equations,

$$x(t_2) = x(t_1) + v(t_1)\delta = 1.000 + (0.000)(.100) = 1.000$$

The acceleration at t_1 is computed from the differential equation itself, i.e., from Eq. [10-22],[1]

$$a(t_1) = -\frac{g}{l} \sin x(t_1) = -10.00 \sin 1.000 = -8.41$$

This acceleration is assumed to be essentially constant over the short time interval between t_1 and t_2. With it, the velocity at the second point may be computed by means of the constant acceler-

---

[1] Note that the acceleration at t_2 is also -8.41 because $x = 1.000$ there, too.

134

ation formula

$$v(t_2) = v(t_1) + a(t_1)\delta = 0.000 + (-8.41)(0.100) = -0.841$$

This value of velocity provides the slope of the solution function at t_2. It also is drawn in Fig. 10-6 as a short, heavy bar. When it is extended to the right a distance of 0.100 (sec), it produces the value of the solution function at t_3,

$$x(t_3) = x(t_2) + v(t_2)\delta = 1.000 + (-0.841)(0.100) = 0.916$$

The values of $a(t_3)$, $v(t_3)$, and $x(t_4)$ may be determined in like manner. They are -7.93, -1.68, and 0.748, respectively. The next set of values, $a(t_4)$, $v(t_4)$, and $x(t_5)$, are -6.80, -2.48, and 0.500. All of these values of position and velocity are shown plotted in Fig. 10-6.

The trouble with this method of solution is the fact that it takes as the almost constant velocity during the interval from t_{i-1} to t_i the velocity at the *beginning* of the interval and uses it to compute the position at the *end* of the interval. Similarly, it employs the acceleration at the beginning of the interval to calculate the velocity at the end. Feynman has shown that a great improvement in accuracy results if the velocity used in all calculations associated with the interval t_{i-1} to t_i is the velocity at the *middle* of the interval, i.e., is $v(t_{i-1} + \frac{1}{2}\delta)$.[1]

According to his procedure, the computational steps are

Start with the values of position and acceleration calculated in the previous interval, i.e.,

$x(t_{i-1})$ and $a(t_{i-1})$,

and the velocity at $t_{i-1} + \frac{1}{2}\delta \equiv t_{i-1}'$,

$v(t_{i-1} + \frac{1}{2}\delta) \equiv v(t_{i-1}')$

which also has been calculated in the previous cycle.

ii Using $v(t_{i-1} + \frac{1}{2}\delta)$, compute $x(t_i)$

$$x(t_i) = x(t_{i-1}) + v(t_{i-1} + \frac{1}{2}\delta)\delta$$

iii Compute the acceleration $a(t_i)$ from the differential equation, Eq. [10-22],

$$a(t_i) = -\frac{g}{l}\sin x(t_i)$$

[1] R. Feynman, R. Leighton, and M. Sands, "The Feynman Lectures on Physics," chap. 9, Addison-Wesley Publishing Company, Inc., Reading, Mass., 1963.

iv   Use $a(t_i)$ to compute the velocity at $t_i + \frac{1}{2}\delta \equiv t_i'$,

$$v(t_i + \tfrac{1}{2}\delta) = v(t_{i-1} + \tfrac{1}{2}\delta) + a(t_i)\delta$$

This is the method that will be used in our computer program to solve Eq. [10-22]. It is very instructive to compute "by hand" the first several solution points, in order to understand exactly how the solution proceeds. This should be done *before* the Fortran program is written; the final computer output should be checked against these hand calculations to ensure that the program has been correctly done.

One small problem in this method still has not been solved: exactly how to *initialize* the computation? We know $x(t_1)$ and $v(t_1)$ from the initial conditions, Eqs. [10-23] and [10-24]. However, according to the preceding discussion, we must begin our computation with $v(t_1 + \tfrac{1}{2}\delta) \equiv v(\tfrac{1}{2}\delta)$, *not* $v(0)$. Feynman resolves this problem by computing $v(\tfrac{1}{2}\delta)$ from the values of v and a at $t_1 = 0$, using the constant-acceleration formula

[10-25] $v(\tfrac{1}{2}\delta) = v(0) + a(0)(\tfrac{1}{2}\delta)$

The value of $a(0)$ is calculated from $x(0)$ by means of the differential equation, Eq. [10-22], as mentioned in step *iii* above.

4. PROGRAMMING   The computational steps outlined in the preceding section may be summarized concisely in the following set of equations

$$x_i = x_{i-1} + v_{i-1}\delta$$

[10-26] $a_i = -\dfrac{g}{l}\sin x_i$

$$v_i = v_{i-1} + a_i\delta$$

In writing Eqs. [10-26], we have used the notation

$x_i \equiv x(t_i)$
$a_i \equiv a(t_i)$
$v_i \equiv v(t_i + \tfrac{1}{2}\delta)$

Note that the velocity v_i is evaluated at a time one-half interval *later* than the time at which x_i and a_i are calculated. This type of notation greatly simplifies the form of the arithmetic statements needed in our Fortran program.

The flow chart of a program to solve Eq. [10-22] subject to the initial conditions

[10-27a] $x(0) = x_0$

and

[10-27b] $v(0) = v_0$

is presented in Fig. 10-7. It computes the position, velocity, and acceleration of the pendulum point by point. It does so in a single DO loop by means of arithmetic statements which are equivalent to Eqs. [10-26].

Before the DO loop is entered, data input and initialization procedures must be performed. Values of x_0, v_0, g, l, and δ must be read in along with n, the number of points desired in the solution. It is convenient to represent each of the system variables x, v, a, and t by a one-dimensional Fortran variable. Let these Fortran variables be designated X, V, A, and T. Each of these arrays must be initialized before the DO loop is begun. The first element of X, X(1), is set equal to the value read in for x_0. T(1)

FIG. 10-7
Flow chart of a Fortran program to solve numerically differential equation [10-22] subject to initial condition Eqs. [10-27]

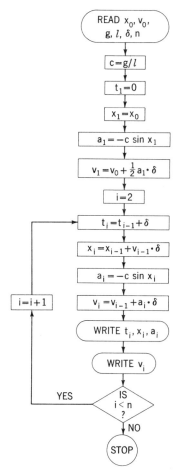

is equated to zero. A(1), according to the second of Eqs. [10-26], is computed by an arithmetic statement like

A(1) = −C*SIN(X(1))

Before this statement is executed, the value of C must be computed as the ratio of g to l.

The value of v_1, or the Fortran variable V(1), is not $v(0)$, but $v(\frac{1}{2}\delta)$. V(1) is computed by means of an arithmetic statement that is equivalent to Eq. [10-25],

V(1) = VO + 0.5*A(1)*DELTA

The DO loop begins with the initial value of its index equal to 2. After arithmetic statements which compute T(I), X(I), A(I), and V(I) have been executed, the computed values of these variables are printed out. It is convenient to present the output in the form of a table similar to Table 4-1. In this kind of table each v_i entry appears between and one line below the x_i and a_i entries. Details of the output and FORMAT statements that produce this type of display are left to the reader.

5. PROCEDURE **a** Set up the apparatus illustrated in Fig. 10-4 to photograph the motion of a simple pendulum. The background against which photographs are taken should be made as nonreflecting as possible. The stroboscope is set off to one side of the pendulum, between it and the camera tripod. If one is available, a Polaroid camera should be used in this experiment, so that a photograph may be produced and measurements made from it all in one laboratory period.

b Adjust the stroboscope to flash at a frequency of 20 to 50 flashes/sec. The camera exposure time should be set to about half of the pendulum period. For a 1-m-long pendulum, the exposure time is about 1 sec; for a 25-cm pendulum, about $\frac{1}{2}$ sec. The stroboscope frequency should be adjusted so that at least 10 to 12 exposures occur during the time the camera shutter is open. It should also be set at some regular value that is easily calibrated, such as 30 Hz. If the stroboscope flashes too rapidly, the photographic images will overlap and make the analysis more difficult.

c Swing the pendulum out to a desired initial displacement angle and hold it there. Note that in the theoretical analysis of the experiment, angles must be expressed in radians. Therefore, a convenient initial value of the amplitude would be 57.3° (1 rad) or some submultiple of it.

d Start the stroboscope flashing at its prescribed frequency. Release the pendulum and open the camera shutter at the same instant. This might require a bit of practice (with an unloaded camera!). Stop the pendulum. Make one additional exposure with the pendulum at rest in a vertical orientation. This is needed as a reference line in the analysis of the photographic data.

e Fasten the photograph to a flat surface. Measure as accurately as possible the angle made by the pendulum with the vertical reference exposure for each of the other flash exposures. Perform the analysis of these data as outlined in Section 2 so that the $t = 0$ position of the modified pendulum data corresponds to an angle of $0°$, i.e., to the equilibrium position. Tabulate the θ-versus-t data in a form similar to Table 10-1. These data will be compared with the theoretical θ-versus-t data obtained from the computer analysis.

f Referring to the flow chart of Fig. 10-7, write a Fortran program to solve Eq. [10-20] subject to the initial conditions indicated in Eqs. [10-27], with $v_0 = 0$. A suitable time increment δ for the computation is 0.02 or 0.05 sec, or even 0.1 sec (with some loss of accuracy). In order to make the time scale of the computed $x(t)$ coincide with that of the $\theta(t)$ obtained experimentally, the $x(t)$ data must be shifted by one-quarter period. That is so because the $\theta(t)$ data begin with $\theta(0) = 0$, while the $x(t)$ data start with $x(0)$ at a maximum. Plot $x(t)$ after shifting the time axis one-quarter period. Superimpose the experimental $\theta(t)$ points on the graph.

g As an interesting extension of this experiment modify the program so that it repeats its computations for a number of different initial displacement angles. This might be done with a GO TO statement which refers to a READ near the beginning of the program. Thus, a new data card with a different value of XO is read after each computation has been completed. From the computed data, determine how the period of a simple pendulum varies with amplitude of vibration. Also determine the variation of period with amplitude experimentally and compare with the computed theoretical results.

EXPERIMENT 11
COUPLED OSCILLATORS

1. INTRODUCTION   In the previous experiment, a simple numerical procedure for the point-by-point solution of an ordinary differential equation was described. The type of equation studied was a second-order equation which could be written in the form

[11-1] $$\frac{d^2x}{dt^2} = f[x(t)]$$

This type of differential equation arises when Newton's second law is applied to physical systems in which conservative one-dimensional forces act. The examples considered in Experiment 10 were the simple harmonic oscillator, the simple pendulum, and an object falling near the surface of the earth.

In this experiment, we are going to consider a related kind of problem in differential equations, the problem of coupled equations. These equations arise from physical situations in which two or more bodies or particles moving under the influence of some external force also interact with one another. The interacting bodies in this experiment take the form of simple pendulums. The interaction force is provided by a light spring that connects them. This is probably the simplest and most straightforward example of coupled systems.

The differential equations that arise from this kind of problem are very similar to Eq. [11-1]. However, because of the interaction between the oscillators, the coordinates of one oscillator become intermixed in the equation of the other.

If we indicate the position of one oscillator by the coordinate x_1, which is measured from its equilibrium position, and the position of the other by x_2, measured from its equilibrium position, the differential equations of motion take the form

[11-2]

$$\frac{d^2x_1}{dt^2} = f[x_1(t), x_2(t)]$$

$$\frac{d^2x_2}{dt^2} = g[x_2(t), x_1(t)]$$

Equations of this kind will be derived for a pair of coupled pendulums in Section 3. The intermixing of the coordinates of the two oscillators is apparent in Eqs. [11-2]. The method of solution of these equations which we will employ is very similar to the procedure that we used in the last experiment. It will be described in Sections 3 and 4.

2. APPARATUS The coupled pendulum apparatus that is used in this experiment is illustrated in Fig. 11-1. It consists of two identical pendulums connected by a light, flexible coupling spring S. Both pendulums consist of a pair of massive iron or steel cylindrical weights fastened to the lower end of a light steel bar about 1 m long. The upper end of each pendulum is drilled through with a smooth $\frac{1}{4}$-in. hole, which is to serve as the pivot axis of that pendulum. It is passed over one of two knife-edge supports, A and B. The distance between knife edges is made equal to the length of the coupling spring, or slightly greater. The pendulums are labeled P_1 and P_2 in the diagram.

Coupled pendulums exhibit an interesting alternation of oscillation. This will be discussed in the next section. The graph presented in Fig. 11-2 shows the displacements of the two pendulums of Fig. 11-1 both plotted as functions of time on a common time axis. These plots were constructed from data produced by a computer program of the type described in Section 4. It is instructive to follow the buildup of oscillation of either pendulum versus time. This may be done experimentally by timing both the period of oscillation of one pendulum and the time required for it to build up to its maximum amplitude and then return to zero amplitude.

FIG. 11-1
Coupled pendulums,
showing coordinates, x_1,
x_2, θ_1, and θ_2

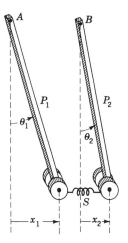

FIG. 11-2
Displacements, $x_1(t)$ and
$x_2(t)$, of two coupled
pendulums

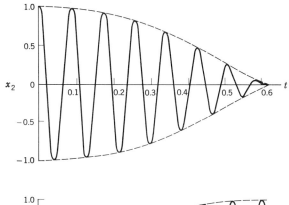

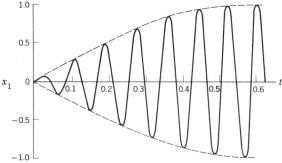

Experiments with coupled oscillators may also be performed on an air track. The air-track apparatus is described in detail in Appendix B. An alternative experiment involving spring-coupled oscillators on an air track is suggested in Section 5. The corresponding differential equations are also discussed there (part h of the procedure section).

3. THEORY When either pendulum makes an angle with vertical which is small (on the order of several degrees), the displacement of the bob of that pendulum from its equilibrium position is very nearly horizontal. This may be clearly seen in Fig. 11-3, the force diagram of one simple pendulum. It is displaced from the vertical equilibrium position by a small angle θ. The differential equation of motion of such a pendulum, previously derived in Experiment 10, in terms of the coordinate θ is

$$[11\text{-}3] \qquad \frac{d^2\theta}{dt^2} = -\frac{g}{l}\sin\theta$$

For small angles θ, of the order of $10°$ or less, the small-angle

143

FIG. 11-3
Force diagram of one
pendulum

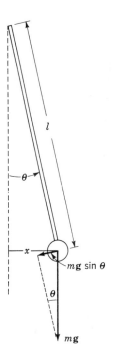

FIG. 11-3
Force diagram of one
pendulum

approximation $\sin \theta = \theta$ may be applied to Eq. [11-3].[1] When this is done, Eq. [11-3] becomes

[11-4] $$\frac{d^2\theta}{dt^2} = -\frac{g}{l}\theta$$

This equation is identical in form to Eq. [10-13], the equation of motion of a simple harmonic oscillator.

The differential equation of motion of the pendulum, Eq. [11-4], may also be written in terms of a coordinate x, which measures the horizontal displacement of the center of the pendulum bob. This coordinate is indicated in Fig. 11-3. There is an approximate relation between x and θ which is valid for small angles (on the order of 10° or less). Referring to the large right triangle in Fig. 11-3, whose hypotenuse is of length l, we have $\sin \theta = x/l$, from the definition of the sine (side opposite θ divided by the hypotenuse). Once again applying the small-angle formula $\sin \theta = \theta$, and solving explicitly for x, we obtain the approximate relation between x and θ

[11-5] $$\theta = \frac{x}{l}$$

[1] For an angle θ of 10°, the values of $\sin \theta$ and θ in radians are, respectively, 0.17365 and 0.17453. Their difference is only about $\frac{1}{2}$ percent. The percentage difference becomes even smaller as θ decreases.

In Eq. [11-5], x and θ are variables; l is a constant. When this equation is differentiated twice with respect to time, it becomes $\ddot{\theta} = \ddot{x}/l$. Using this equation and Eq. [11-5] to eliminate $\ddot{\theta}$ and θ in Eq. [11-4], we obtain the differential equation of motion of the pendulum in terms of the coordinate x,

[11-6] $\quad \dfrac{d^2x}{dt^2} = -\dfrac{g}{l}x$

Apart from the coefficient of its x term, this equation is identical to Eq. [10-11].

Two uncoupled pendulum oscillators obey differential equations like Eq. [11-6]. If we denote the horizontal displacement of the two pendulum bobs by x_1 and x_2, we may write these equations as

[11-7] $\quad \dfrac{d^2x_1}{dt^2} = -\dfrac{g}{l}x_1 \quad$ and $\quad \dfrac{d^2x_2}{dt^2} = -\dfrac{g}{l}x_2$

The coupling force causes Eqs. [11-7] to be transformed into equations of the type of Eqs. [11-2]. The way this occurs may be deduced from a consideration of the forces applied by the spring to the two pendulum bobs. Figure 11-4 is a close-up of the spring and the two bobs, showing the force that the spring exerts on each of them. The displacement of the bob at the right, x_2, is assumed greater than that of the bob at the left, x_1. Therefore, the spring is in tension. It pulls toward the right on bob 1 and toward the left on bob 2. The net extension of the spring is $x_2 - x_1$. By Hooke's law, the magnitude of the spring force is then

[11-8] $\quad F_s = k(x_2 - x_1)$

The x components of the spring force on the two bobs are thus

FIG. 11-4
Force diagram of the two coupled pendulums

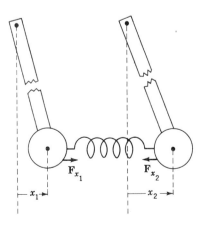

[11-9]
$$F_{x_1} = +k(x_2 - x_1)$$
$$F_{x_2} = -k(x_2 - x_1)$$

When Eqs. [11-9] are substituted along with the x component of the weight force into the Newton's second law equation for each oscillator, the following equations result:

[11-10] $\quad k(x_2 - x_1) - \dfrac{mg}{l}x_1 = m\ddot{x}_1$

[11-11] $\quad -k(x_2 - x_1) - \dfrac{mg}{l}x_2 = m\ddot{x}_2$

After we divide both equations through by m and simplify them, we obtain a pair of equations of the form of Eqs. [11-2],

[11-12] $\quad \ddot{x}_1 = -\left(\dfrac{g}{l} + \dfrac{k}{m}\right)x_1 + \dfrac{k}{m}x_2$

[11-13] $\quad \ddot{x}_2 = -\left(\dfrac{g}{l} + \dfrac{k}{m}\right)x_2 + \dfrac{k}{m}x_1$

We may rewrite the coefficients of the terms on the right-hand sides of Eqs. [11-12] and [11-13] more simply as

[11-14] $\quad A = -\left(\dfrac{g}{l} + \dfrac{k}{m}\right) \quad$ and $\quad B = \dfrac{k}{m}$

With these coefficients, the differential equations of motion of the system, Eqs. [11-12] and [11-13], become

[11-15] $\quad \ddot{x}_1 = -Ax_1 + Bx_2$

and

[11-16] $\quad \ddot{x}_2 = -Ax_2 + Bx_1$

The method which will be used to solve these equations is very similar to the procedure employed in the preceding experiment. It may be summarized in the following sequence of operations

i   Start with positions and accelerations evaluated at the time $t_{i-1} = t_i - \delta$,

$x_1(t_{i-1}) \qquad x_2(t_{i-1}) \qquad a_1(t_{i-1}) \qquad a_2(t_{i-1})$

and velocities at $t_{i-1}' = t_i - \tfrac{1}{2}\delta$,

$v_1(t_{i-1}') \qquad$ and $\qquad v_2(t_{i-1}')$

all of which have been computed in the previous cycle.

146

ii   Compute $x_1(t_i)$ and $x_2(t_i)$ using these velocities,

$$x_1(t_i) = x_1(t_{i-1}) + v_1(t_{i-1}')\delta$$
$$x_2(t_i) = x_2(t_{i-1}) + v_2(t_{i-1}')\delta$$

iii Compute the acceleration components from the differential equations [11-15] and [11-16] and the displacements just found in step ii,

$$a_1(t_i) = -Ax_1(t_i) + Bx_2(t_i)$$
$$a_2(t_i) = -Ax_2(t_i) + Bx_1(t_i)$$

iv   Compute the velocity components at the middle of the next interval,

$$v_1(t_i') = v_1(t_{i-1}') + a_1(t_i)\delta$$
$$v_2(t_i') = v_2(t_{i-1}') + a_2(t_i)\delta$$

Note that the times t_i' and t_{i-1}' at which velocities are evaluated are different from t_i and t_{i-1}, at which positions and accelerations are computed.

When the masses of the two pendulums are not the same, the differential equations which describe their motion must be modified to take this fact into account. Let the two masses be denoted by m_1 and m_2. A derivation identical to the one that led to Eqs. [11-12] and [11-13] in this case produces the set of differential equations

[11-17] $$\ddot{x}_1 = -\left(\frac{g}{l} + \frac{k}{m_1}\right)x_1 + \frac{k}{m_1}x_2$$

and

[11-18] $$\ddot{x}_2 = -\left(\frac{g}{l} + \frac{k}{m_2}\right)x_2 + \frac{k}{m_2}x_1$$

Replacing the coefficients of the terms on the right-hand sides of Eqs. [11-17] and [11-18] by the letters $A, B, C,$ and D, we obtain the equations

[11-19] $$\ddot{x}_1 = -Ax_1 + Bx_2$$

and

[11-20] $$\ddot{x}_2 = -Cx_2 + Dx_1$$

These are the differential equations whose solutions mathematically describe the motion of the system of two pendulums of different mass. The solution procedure is the same as previously outlined for the two identical pendulums.

4. PROGRAMMING   Figure 11-5 is the flow chart of a Fortran program to solve the system of differential equations of motion of coupled pendulums,

147

FIG. 11-5
Flow chart of a program
to solve the coupled
pendulum differential
equations, Eqs. [11-12]
and [11-13]

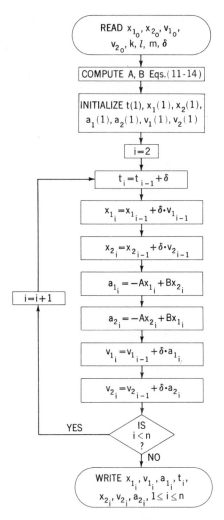

FIG. 11-5
Flow chart of a program to solve the coupled pendulum differential equations, Eqs. [11-12] and [11-13]

Eqs. [11-12] and [11-13]. It is very similar to the program used to solve the simple pendulum differential equation in Experiment 10.

The program begins by reading in values of the system parameters m, k, l, and δ, and initial values of positions (x_{1_0} and x_{2_0}) and velocities (v_{1_0} and v_{2_0}). Then, various initializing procedures are performed. First of all, the coefficients A and B are computed using Eqs. [11-14]. The initial accelerations (at $t = 0$) and velocities (at $t = \frac{1}{2}\delta$) must then be computed in appropriate arithmetic statements, as discussed below.

Each value of position, velocity, and acceleration of either pendulum at each instant of time, t_i, is represented by one element of a one-dimensional subscripted variable. The initial displacements of the two pendulums are read into the first elements of the Fortran variables X1 and X2. The initial accelerations of the two pendulums, $a_1(0)$ and $a_2(0)$, are then computed by means of Eqs. [11-15] and [11-16] from these initial position values. The "initial" velocities are computed next by means of the equations

[11-21] $v_1(\tfrac{1}{2}\delta) = v_{1_0} + a_1(0)(\tfrac{1}{2}\delta)$

and

[11-22] $v_2(\tfrac{1}{2}\delta) = v_{2_0} + a_2(0)(\tfrac{1}{2}\delta)$

The initial velocities which appear in Eqs. [11-21] and [11-22] are really the values of the velocities at $t = \tfrac{1}{2}\delta$, as previously discussed. These equations are analogous to Eq. [10-25], which was employed to initialize the velocity in the numerical solution of the simple pendulum differential equation.

The system of coupled oscillators has the position coordinates, velocities, and accelerations of both oscillators inextricably mixed together in all of its equations. For instance, in order to find the acceleration of either pendulum at any given instant of time, we must have knowledge of the positions of *both* pendulums at that instant. This is required by Eqs. [11-12] and [11-13].

The velocities of the two oscillators are computed from these accelerations, and thus the velocity of each depends on the position coordinates of *both* oscillators. These velocities are then used to compute the positions of the two pendulums in the following time interval, and so forth. The program illustrated in Fig. 11-5 takes account of these relationships, and in each pass through the computational DO loop computes all of the system variables x_1, x_2, v_1, v_2, a_1, and a_2.

5. PROCEDURE   **a**   Suspend the two pendulums from knife-edge supports which are separated by a distance equal to the relaxed length of the coupling spring. Connect the coupling spring, which should be a light, flexible spiral spring, between the two pendulums. See Fig. 11-1.

b   Hold one pendulum still and displace the other through an angle of several degrees, thereby stretching the coupling spring. Do not displace the pendulum through so large an angle that excessive stretching of the spring occurs. Release the displaced pendulum, and at the same time, without imparting any velocity to it, also release the still pendulum.

c   Time the period of vibration of the oscillating pendulum. Notice that very soon after the pendulums are released, the initially motionless pendulum begins to vibrate. Eventually it acquires the amplitude

149

of oscillation originally possessed by the other pendulum. At this moment, the one initially vibrating with large amplitude has come to a complete (or nearly complete) stop. Determine the time required for this transfer of vibrational energy. Do so by starting a stopwatch at the instant one pendulum has stopped vibrating and stopping the stopwatch after that pendulum has acquired its maximum amplitude and again come to a stop. This measurement requires a good deal of patience on the part of the experimenter, due to the complicated motions of the two pendulums, as does the determination of the period of either pendulum alone.

d Determine the spring constant k in the following manner. Suspend the spring from a horizontal support arm. Attach a weight hanger to the lower end of the spring. Determine the elongation per increment of added weight. This is the spring constant.

e Refer to the flow chart of Fig. 11-5. Write a Fortran program to compute the position, velocity, and acceleration of each pendulum by the pointwise solution of Eqs. [11-12] and [11-13], as outlined in Section 3. The program should begin by reading in values of the system parameters m, k, and l, and of the initial displacement and velocity of each pendulum. An appropriate set of initial conditions for this problem is

$$x_1(0) = 0 \qquad \dot{x}_1(0) = 0$$
$$x_2(0) = x_0 \qquad \dot{x}_2(0) = 0$$

The value read in for x_0 should be the initial displacement of pendulum 2. The pointwise solution procedure is indicated in the flow chart.

f Using the results computed in step e, make plots of x_1 and x_2 versus time, similar to Fig. 11-2. The time scale (abscissa) of both graphs should be the same, and one graph should be plotted above the other, as in the figure. From successive points at which the oscillations of one pendulum decrease to zero amplitude determine the time required for the transfer of energy from one pendulum to the other. Also determine the period of vibration of each pendulum. Compare these values with the experimental values found in step c.

g As an optional extension of the experiment, use bobs of different masses. Perform steps a through f for the modified coupled-oscillator system. The computer program required in step e should carry out a numerical solution of Eqs. [11-19] and [11-20] for $x_1(t)$ and $x_2(t)$.

h An alternative version of the coupled oscillator experiment makes use of the linear air-track apparatus described in Appendix B. Figure 11-6 illustrates an air-track experiment in which two gliders of unequal masses (m_1 and m_2) are arranged to function as coupled oscillators. Two identical springs (of force constant k) connect the gliders to the air-track apparatus, and a weaker spring (of force constant k') con-

FIG. 11-6
Coupled oscillator experiment performed on a
linear air track. The
gliders have unequal
masses m_1 and m_2.

nects them to each other. The differential equations of motion of this system may be shown to be

[11-23] $\ddot{x}_1 = -\left(\dfrac{k}{m_1} + \dfrac{k'}{m_1}\right)x_1 + \dfrac{k'}{m_1}x_2$

and

[11-24] $\ddot{x}_2 = -\left(\dfrac{k}{m_2} + \dfrac{k'}{m_2}\right)x_2 + \dfrac{k'}{m_2}x_1$

Verify these equations. Repeat steps b through f for this system.

EXPERIMENT 12
LEAST-SQUARES ANALYSIS—
THE PHYSICAL PENDULUM

1. INTRODUCTION   The least-squares method of fitting experimental data to a theoretical curve is one of the most frequently used data analysis procedures in experimental science and engineering. It is most commonly applied to data which, according to theory, should follow a straight line graph. These data obey an equation of the form

[12-1] $y = ax + b$

In Eq. [12-1] the constants a and b are the slope and y intercept of the straight line. As its name suggests, the least-squares method minimizes the sum of the squares of the deviations of the experimental points from a "best" straight line drawn through them. In doing so, it supplies the values of the parameters a and b of that best line.

In the most elementary version of the method, the one that we will make use of in this experiment, one coordinate of each point is assumed to be entirely free from error. Let us choose it to be the x coordinate. Then, the entire experimental error in each datum point (x_i, y_i) resides in y_i. Figure 12-1 shows a set of experimental data points through which the least-squares best straight line has been drawn. The difference between a measured y_i and the height of the line at $x = x_i$ is called the deviation δ_i. By Eq. [12-1], the height (ordinate) of the line at $x = x_i$ must be $ax_i + b$. Thus, the deviation of the ith point from the least-squares best line is

[12-2] $\delta_i = y_i - ax_i - b$

The least-squares technique has its basis in statistical theory. The procedure that we are describing applies strictly to y_i data whose deviations from the best straight line follow a normal distribution. In this case, the parameters a and b of that best line, which minimizes the squares of the deviations δ_i, are the statistically *most probable* values of a and b. The best straight line in

153

FIG. 12-1
Experimental data fit by
best straight line in the
least squares sense. The
deviation of the ith point
from the line is indicated.

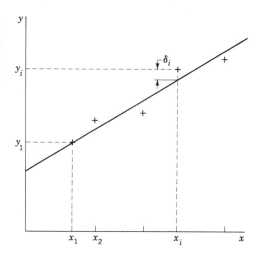

the least-squares sense is the line *defined* by the relation

[12-3] $$\sum_{i=1}^{n} \delta_i{}^2 = \text{minimum}$$

An example of the application of the least-squares technique
in analysis of experimental data comes from Experiment 8. In
the determination of Young's modulus, the change in reading of
the scale of an optical lever per increment of applied load had
to be calculated from scale-reading-versus-load data. In that
experiment, the desired ratio was computed by means of the
method of differences. This ratio might also be determined with
the aid of the least-squares method, as explained below.

In Fig. 12-2, scale-reading data are plotted against values of
applied load; the data very closely follow a straight line. The load
masses are assumed to be entirely accurate. Any errors that arise
in the determination are considered to be caused by the scale-
reading data alone. These are the conditions required for an
elementary least-squares analysis.

If the slope of the best straight line through these data is determined by a least-squares procedure, it will provide a value for
the ratio required in the calculation of Young's modulus. The
intercept of this line in Fig. 12-2 happens to be a negative quantity. It is related to the initial mirror setting of the optical lever
and has no particular experimental significance in the determination. The intercept need not even be computed. In fact, the
origin might equally well be located at the first experimental
point, as indicated by the dashed axes in Fig. 12-2, and the
intercept could thereby be eliminated.

FIG. 12-2
Telescope scale reading
versus applied load, from
the optical lever of
Young's modulus
experiment

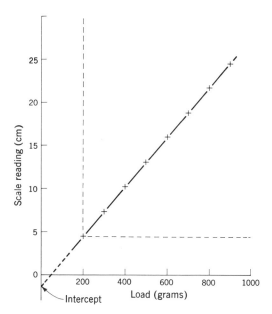

FIG. 12-2
Telescope scale reading
versus applied load, from
the optical lever of
Young's modulus
experiment

Clearly, not all pairs of measurable physical quantities are linearly related to one another. However, by suitably modifying the variables that represent two nonlinearly related physical quantities, it is often possible to cast them into an equation of the form of Eq. [12-1]. If this can be done, it is then possible to perform the elementary kind of least-squares analysis we have been discussing. In the remainder of this section, we will consider several examples of this kind.

The period of a simple pendulum varies as the square root of its length according to the formula

[12-4] $$T = 2\pi\sqrt{\frac{l}{g}}$$

when its amplitude of vibration is held to small angles (about $10°$ or less, as discussed in Experiment 1). We may collect together all constant factors in Eq. [12-4] and rewrite it as

[12-5] $$T = \frac{2\pi}{g^{1/2}}\, l^{1/2}$$

If we define x as $l^{1/2}$ and y as T, we succeed in converting Eq. [12-5] into the form $y = ax + b$. In this equation, the intercept b is zero, and the slope a is equal to the constant factor $2\pi/g^{1/2}$. If the x_i-versus-y_i data are subjected to a least-squares analysis, the gravitational acceleration g may be computed from the slope

155

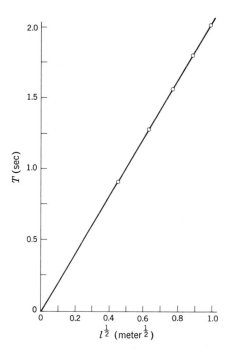

FIG. 12-3
Linearized graph of
simple pendulum data:
T **versus** $l^{\frac{1}{2}}$

of the best-fitting line. Details of this analysis are presented in Section 3. A graph of T versus $l^{1/2}$ for the simple pendulum data of Experiment 1 appears in Fig. 12-3.

A slightly more refined version of this kind of analysis may be performed with period-versus-length data obtained with a physical pendulum. A discussion of the least-squares analysis of data taken with a vibrating thin-rod pendulum will be given in Section 3. If the period of such a pendulum is measured for different pivot axes, each of which is at a different distance from the center of gravity of the pendulum, the period-versus-distance data will turn out to be markedly nonlinear. However, as will be shown, it is again possible to convert the two variables into a linear form suitable for our least-squares procedures.

2. APPARATUS   The physical pendulum apparatus used in this experiment is illustrated in Fig. 12-4. It consists of a rectangular-cross-section steel bar in which $\frac{1}{4}$-in. holes have been drilled at intervals from the center of the bar to one of its ends. For the sake of symmetry, it is well to use a bar with holes drilled at equal distances on either side of its center. This will ensure that the center of gravity of the bar is located at its geometrical center, and that its moment

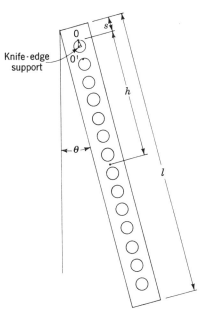

FIG. 12-4
Experimental physical
pendulum

Knife-edge
support

of inertia is approximated closely by the thin-rod formula

[12-7] $I_c = \frac{1}{12}ml^2$

This is the formula for the moment of inertia of a thin rod about an axis that passes through the center of gravity of the rod at right angles to the plane of the rod.

The bar is suspended from a knife-edge pivot which passes through one of the drilled holes. The distance from pivot to center of gravity is labeled h in Fig. 12-4. The length of the rod is indicated by l. The rod is shown in an instantaneous orientation in which its axis makes an angle θ with vertical.

An alternative version of the experiment makes use of a simple pendulum, which consists of a small, heavy bob tied to the lower end of a light cord. The cord is fastened at its upper end to a firm support. The distance from support to the center of the spherical bob is indicated in Fig. 12-5 by l. The cord is illustrated at that instant when it also makes an angle θ with vertical. In order that Eq. [12-4] accurately represent the period of the pendulum, it is necessary that its vibration amplitude be limited to angles θ no larger than $10°$, as previously discussed.

3. THEORY   The least-squares analysis that is described in this section applies to a set of data in which there is no error at all in the x_i coordinates. Furthermore, in order to keep the mathematical treatment

FIG. 12-5
Simple pendulum

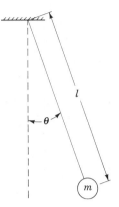

simple, the data will be assumed to conform to a theoretical equation of the form

[12-8] $y = ax$

Equation [12-8] is, of course, the equation of a straight line that passes through the origin. Even though this condition is somewhat restrictive, there are many practical applications for a linear analysis of this kind. Each of the examples previously discussed in this experiment could be represented by a formula like Eq. [12-8] by algebraically modifying one or both of the experimental variables.

According to Eqs. [12-2] and [12-3] we must minimize the following expression:

[12-9] $$\sum \delta_i^2 = \sum_{i=1}^{n} (y_i - ax_i)^2$$

Specifically, what we must do is: find the slope a of that particular straight line drawn through the origin which causes the sum of the squares of the deviations to be a minimum. Equation [12-9] has only one parameter that must be determined. It is a, the slope of the line. The minimization problem is a standard exercise in differential calculus. Since all of the (x_i, y_i) data are mathematical constants (they are just experimental *numbers* insofar as this calculation is concerned), the summation in Eq. [12-9] is a function of one variable (parameter). To find the value of a that minimizes the function, we differentiate the sum on the right-hand side of Eq. [12-9], set the derivative equal to zero, and solve for a.

Differentiating Eq. [12-9] and setting the result equal to zero, we obtain

$$0 = \frac{d}{da} \sum_{i=1}^{n} (y_i - ax_i)^2 = 2 \sum_{i=1}^{n} (y_i - ax_i)(-x_i)$$

Factoring out a and rewriting this equation as two sums, we have

$$a \sum_{i=1}^{n} x_i^2 - \sum_{i=1}^{n} x_i y_i = 0$$

Finally, solving the above equation explicitly for a, we obtain the least-squares expression for the slope of the straight line that best fits the (x_i, y_i) data,

$$[12\text{-}10] \quad a = \frac{\displaystyle\sum_{i=1}^{n} x_i y_i}{\displaystyle\sum_{i=1}^{n} x_i^2}$$

A more general least-squares equation may be derived for the slope of the best-fitting straight line. It is useful whether or not the data obey Eq. [12-8], i.e., whether or not the line intersects the origin. The derivation depends on the fact that the least-squares best line *always* passes through the *center of gravity* or *centroid* of the experimental data points.[1] Let the centroid be denoted by $(\bar{x}, \bar{y})$. It is defined by the equations (which are analogous to those employed to calculate the centroid of a collection of point masses)

$$[12\text{-}11] \quad \bar{x} = \frac{1}{n} \sum_{i=1}^{n} x_i \quad \text{and} \quad \bar{y} = \frac{1}{n} \sum_{i=1}^{n} y_i$$

The derivation also requires that the sum of squares of deviations be minimized with respect to a. When this is done, and the result is combined with that given above, the slope is found to be

$$[12\text{-}12] \quad a = \frac{n \displaystyle\sum_{i=1}^{n} x_i y_i - \displaystyle\sum_{i=1}^{n} x_i \displaystyle\sum_{i=1}^{n} y_i}{n \displaystyle\sum_{i=1}^{n} x_i^2 - \left(\displaystyle\sum_{i=1}^{n} x_i\right)^2}$$

Let us reconsider the treatment of the simple pendulum data of Experiment 1. Making use of the period-versus-length data that

[1] From Eqs. [12-2] and [12-3], the sum that must be minimized in the general case is $\Sigma(y_i - ax_i - b)^2$. When it is differentiated with respect to the parameter b and the result is set equal to zero, the equation $\Sigma y_i - a\Sigma x_i - nb = 0$ is obtained. From Eq. [12-11], the sums in this equation, Σx_i and Σy_i, are equal to $n\bar{y}$ and $n\bar{x}$. Therefore, the equation is equivalent to $n\bar{y} - na\bar{x} - nb = 0$, or $\bar{y} = a\bar{x} + b$. The latter equation tells us that the point $(\bar{x}, \bar{y})$ lies on the line whose slope and intercept are a and b, respectively. Thus the best straight line in the least-squares sense does indeed pass through the centroid.

TABLE 12-1
Least-squares Analysis of
Simple Pendulum Data

$x_i{}^2 = $ length l_i m	$y_i = $ period T_i sec	$x_i = \sqrt{l_i}$ $m^{1/2}$	$x_i y_i$ $m^{1/2} \cdot$ sec
0.200	0.898	0.4472	0.4016
0.400	1.269	0.6325	0.8026
0.600	1.554	0.7746	1.2037
0.800	1.794	0.8944	1.6046
1.000	2.007	1.0000	2.0070
$\Sigma x_i{}^2$: 3.000	Σy_i: 7.522	Σx_i: 3.7487	$\Sigma x_i y_i$: 6.0195

were presented there, we may perform a least-squares analysis of the type described in Section 1. The computations required are summarized in Tables 12-1 and 12-2. The slope of the best line is determined with the aid of Eq. [12-10] or Eq. [12-12]. From Eq. [12-5], this slope must be equal to $(2\pi/g^{1/2})$. Therefore, the gravitational acceleration g must be given by the relation

[12-13] $$g = \frac{4\pi^2}{a^2}$$

Equation [12-13] will be used in the computer program of Section 4 to solve for g.

We have retained one more figure in these tables than is warranted by the accuracy of our data. This is always a good idea when carrying through calculations. According to the discussion of Experiment 1, Eq. [12-5] is only valid to about 0.2 percent anyway when $\theta = 10°$. Therefore, we ought to report our final result as $g = 9.80_6$ m/sec^2, or even better, as 9.81 m/sec^2.

Before going on to describe the least-squares analysis of the physical pendulum data, let us derive an expression for its period as a function of the distance between pivot and center of gravity. We will apply this expression to the thin-rod pendulum. With a bit of algebraic manipulation we will obtain an equation for the period in the form of Eq. [12-8]. Using it, we may perform a least-squares treatment of the experimental data to obtain a value of the gravitational constant g.

A force diagram of a physical pendulum of arbitrary shape is given in Fig. 12-6. The weight force $\mathbf{w} = m\mathbf{g}$ is shown resolved into components parallel and perpendicular to the axis of the pendulum, i.e., the line that connects the pivot point and center of gravity. Only the perpendicular component of $\mathbf{w}$ gives rise to

TABLE 12-2
Results of Least-squares
Analysis

Slope of $y(x)$ graph:	2.0065 sec/m$^{1/2}$
Gravitational constant:	9.806 m/sec^2

FIG. 12-6
Force diagram of a
physical pendulum

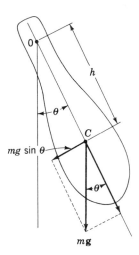

FIG. 12-6
Force diagram of a
physical pendulum

a torque about the pivot axis 0 (an axis through point 0, the pivot point, and perpendicular to the pendulum axis). The torque is the product of this force component and its lever arm h,

[12-14] $\tau = -mgh \sin \theta = -mgh\theta$

In writing Eq. [12-14], we have made use of the small-angle approximation $\theta \approx \sin \theta$, which is accurate to about 0.5 percent when θ is less than $10°$.[1] The minus sign refers to the fact that the torque is negative (clockwise) when the angle θ is positive (counterclockwise) and vice versa.

The rotational analog of Newton's second law, $\Sigma F = ma$, is

[12-15] $\Sigma \tau = I_0 \alpha$

It relates the torques acting on a body pivoted about axis 0 to the angular acceleration α of the body and its moment of inertia I_0 about that axis. Substituting the torque expression of Eq. [12-14] into Eq. [12-15], and rewriting the angular acceleration α as $\ddot{\theta}$, we obtain the differential equation of motion of a physical pendulum,

[12-16] $\ddot{\theta} = -\dfrac{mgh}{I_0}\theta$

[1] Had we not made the small angle approximation in writing Eq. [12-14], we would have obtained a nonlinear differential equation like Eq. [10-20]. When the amplitude of vibration of the pendulum is restricted to angles less than about $10°$, the approximate solution of this equation yields the same expression for the period that we obtain in Eq. [12-18]. For larger amplitudes, correction terms like the one given in Eq. [1-6] must be added to this expression. Equation [12-18] is accurate to within 0.2 percent when θ is kept less than $10°$.

Equation [12-16] has exactly the same form as the simple harmonic motion differential equation, Eq. [10-11]. Therefore, Eq. [12-16], must also have a sinusoidal time-dependent solution. In the harmonic motion case, the period of oscillation is related to the constant of proportionality between $\ddot{x}$ and $-x$, k/m, by

[12-17]　$T = 2\pi \sqrt{\dfrac{m}{k}}$

Thus, by analogy, the period of a physical pendulum must be given by the relation

[12-18]　$T = 2\pi \sqrt{\dfrac{I_0}{mgh}}$

In Eq. [12-18], I_0 is the moment of inertia of the physical pendulum about the pivot axis 0, and h is the distance between 0 and the center of gravity, C, in Fig. 12-6. For the case of the thin-rod pendulum illustrated in Fig. 12-4, I_0 may be conveniently derived from I_c, the moment of inertia about an axis through the center of gravity of the rod. This may be done with the aid of the parallel-axis theorem. This theorem states that the moment of inertia of a body about any axis is equal to the moment of inertia about a parallel axis through its center of gravity plus the product of its mass and the square of the distance between axes. Applied to the thin-rod pendulum, it yields the following expression for I_0:

[12-19]　$I_0 = \frac{1}{12}ml^2 + mh^2$

When this result is substituted into Eq. [12-18], and the factor m, which is common to all terms inside the radical, is canceled, the period of the thin-rod pendulum is found to be

[12-20]　$T = 2\pi \sqrt{\dfrac{\frac{1}{12}l^2 + h^2}{gh}}$

This result is a distinctly nonlinear function of h. A plot of T versus h is rather complicated and does not lend itself readily to further analysis. This plot is given in Fig. 12-7. The solid line is the theoretical graph calculated from Eq. [12-20]. The experimental points follow it rather closely. However, just how one might go about analyzing a plot of this kind to obtain a value of the gravitational constant is not clear.

Equation [12-20] may be converted into a form suitable for least-squares analysis. This may be done by defining a somewhat

162

FIG. 12-7
Period versus h for the
physical pendulum of
Fig. 12-4

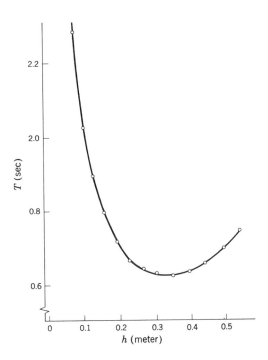

complicated expression for the variable x. If we write x as

$$[12\text{-}21] \quad x = \sqrt{h + \frac{l^2}{12h}}$$

Eq. [12-20] may be written

$$[12\text{-}22] \quad T = \frac{2\pi}{g^{1/2}} x$$

This equation has the form of Eq. [12-8] and may be subjected to a least-squares analysis. The value computed for the slope in this analysis is related to g; g may be computed from it by means of Eq. [12-13].

Implicit in the analysis just described is the assumption of complete accuracy of the x_i data. We ought at this point to consider briefly the validity of this assumption. The variable x is composed of the length measurements l and h. Both of these lengths can be measured to an accuracy of better than 1 mm, or on the order of several tenths of one percent at most. Furthermore, x is computed from them by taking the square root of an expression in which they appear to the first power (or negative first power). As we learned in Experiment 2, this has the effect of *halving* the percent error that either contributes to the percent error in x. Thus, the error in x is at most on the order of 0.1 percent.

The dependent variable in Eq. [12-22] is the period of vibration

T of the pendulum. The measurement of T is subject to various timing errors, principally due to reaction times of the observer in starting and stopping a stopwatch. This error can be minimized by increasing the number of vibrations timed. Nevertheless, the error in T is generally a much larger percentage error than that in x, especially when a relatively small number of vibrations are timed. Therefore, our basic assumption that all of the experimental error in this determination is contained in the y_i variable is reasonably well justified.

4. PROGRAMMING The Fortran program required for the least-squares analysis in this experiment is very straightforward. Apart from data input and output statements, it consists primarily of a single computational DO loop. The DO loop performs the linearization of the x_i data and computes either two or four sums for Eq. [12-10] or [12-12]. The linearization occurs in the flow chart block marked COMPUTE X(I). In the simple pendulum case, the square roots of all pendulum lengths must be taken. After the square roots are computed, the results are returned to the X array. This may be accomplished with an arithmetic statement such as

X(I) = SQRT(X(I))

In the case of the physical pendulum data, the linearization procedure is slightly more complicated. It requires a Fortran arithmetic statement equivalent to Eq. [12-21]. In both cases, the Y array contains the pendulum period data. These data are read in and stored in Y by means of a READ statement located near the beginning of the program.

The two sums required by Eq. [12-10] to evaluate the slope of the best-fitting line are indicated in the flow chart by the variables SUM1 and SUM2. Before the computational loop is started, it is necessary to clear the storage locations in which the sums are to be collected. The terms of the sum $\Sigma x_i y_i$ are added together and the sum is accumulated in SUM1. The sum Σx_i^2 is formed and stored in SUM2.

After its component summations have been computed, the slope of the best line is calculated in an arithmetic statement which is equivalent to Eq. [12-10] or Eq. [12-12]. The gravitational constant g is computed from this value of the slope in an arithmetic statement that corresponds to Eq. [12-13].

It is convenient to include in the program provision for printing out first the unlinearized (x_i, y_i) data, and then, after linearization,

FIG. 12-8
Flow chart of a program
to perform a least-squares
analysis of simple pen-
dulum or physical pendu-
lum data. The length l
is needed only in the
physical pendulum case;
it is the total length of
the pendulum rod.

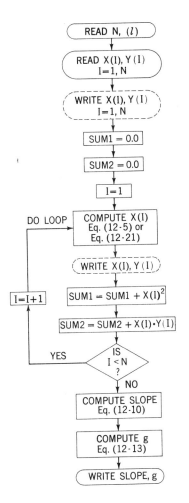

the modified (x_i, y_i) data. The corresponding output statements are indicated in the dashed-line boxes in the flow chart.

5. PROCEDURE   **a**   If the physical pendulum version of this experiment is being performed, first accurately measure and record the total length of the
pendulum, l. Then measure and record the distance between one end
of the pendulum and point O of the drilled hole nearest this end. This
distance is labeled s in Fig. 12-4.

b   Compute the distance between the center of mass of the rod and
the pivot axis O. This is done by subtracting from $\frac{1}{2}l$ the length s
measured in part a. The result is the length indicated by h in Fig.
12.4. Repeat steps a and b for all holes on one side of the center
of mass. These measurements should be made to the nearest tenth
of a millimeter (which must be estimated).

c Set the pendulum swinging in a vertical plane perpendicular to the pivot axis O. There should be no "wobbling" of the pendulum in and out of this plane. The maximum amplitude of oscillation should be restricted to angles less than $10°$ during the entire experiment.

d Time 10 complete vibrations of the pendulum; record T. Suspend the pendulum on the knife-edge support so that its pivot axis is at the top of the next hole, i.e., at the point labeled O' in Fig. 12-4. Time 10 vibrations; record T. Repeat the determination of T in this manner for all holes on one side of the center of mass.

e Referring to the flow chart of Fig. 12-8, write a Fortran program to perform a least-squares data analysis, as discussed in Sections 3 and 4. Determine the slope of the best fitting straight line, and from it the gravitational acceleration constant g.

f Plot a graph of the linearized (x_i, y_i) data points. Draw on the graph the straight line that visually seems to provide the best fit to the data points. Determine the slope of this line. Note that, because it must conform to Eq. [12-22], the line must pass through the origin. Now, superimpose on the graph the best fitting line from the least-squares analysis, i.e., the line through the origin which has the slope computed from Eq. [12-10] or Eq. [12-12]. Compare the slopes of the two lines and the values of g that they produce.

g If the simple pendulum version of the experiment is being performed, start by suspending the pendulum from a rigid support with the distance between bob (center) and support very close to 20 cm. Measure and record the exact value of this distance. Time 10 oscillations; record T. Readjust the length of the pendulum to 30 cm and repeat. Continue until the pendulum is too long to swing freely. If necessary, use a length increment of 5 cm, especially at low values of l. Obtain in this way at least 10 pairs of (l_i, T_i) data.

h Write a Fortran program to compute the slope of the best-fitting straight line and the gravitational constant g. Perform the same graphical analysis with these simple pendulum data as is suggested in step f. Compare values of slope and g from the computer output with the values obtained from the line drawn "by eye."

i As an alternative data analysis procedure, write a computer program to determine Young's modulus from data obtained in Experiment 8. The program should perform a least-squares analysis of optical lever data to determine the ratio of scale reading change per unit of added load mass. In order to apply the simple least-squares technique discussed in this experiment, the data must be converted into the form of Eq. [12-8]. This is most easily done by shifting the coordinate axes so that the origin coincides with the first datum point.

Fortran equations of the kind

X(I) = X(I) − X(1)
Y(I) = Y(I) − Y(1)

will accomplish the transformation. Here, Y is used to represent optical-lever scale readings, and X, load mass. Compare the value of Young's modulus obtained in this manner with the value previously found in Experiment 8.

j   As another alternative application of the least-squares technique, write a program to compute g from the instantaneous velocity-versus-time data obtained in Experiment 4.

EXPERIMENT 13
CALORIMETRY—NUMERICAL INTEGRATION

1. INTRODUCTION   The modern view of heat is that it is simply another form of energy. This concept of the equivalence between heat and mechanical energy was convincingly demonstrated over one-hundred years ago by Sir James Prescott Joule in his famous paddle-wheel experiment.[1] The practical unit of heat energy, the calorie, is in fact today defined in terms of the mechanical energy unit named after Joule. One calorie is defined as 4.186 joules. The older definition specified the calorie to be the amount of heat that must be added to 1 g of water initially at 14.5 °C to raise its temperature to 15.5 °C. The latter definition is still useful. It does provide a physical interpretation of the calorie as a unit of heat.

Whenever a quantity of matter undergoes a change of temperature, there is an accompanying transfer of energy, i.e., heat, to it or from it. If its temperature increases, energy must have been added to it. If its temperature drops, energy must have been extracted from it. The physical quantity that relates the heat added (or extracted) from a given mass of a substance to the temperature change that it experiences is called the *specific heat* of the substance. The specific heat is usually denoted by the letter c and is defined by the equation

[13-1] $$\Delta Q = mc\,\Delta T$$

In Eq. [13-1], m is the mass of material undergoing the change in temperature ΔT, and ΔQ is the heat added to it (if ΔT is positive, or extracted from it if ΔT is negative). The usual units of c are calories per gram per degree Celcius. For most materials within a range of several tens of degrees either side of room temperature, the specific heat is approximately constant. For example, the specific heat of water between 0 and 100°C is 1.00 cal/g·°C within 1 percent.

---

[1] In the paddle-wheel experiment, the mechanical energy loss of a falling weight is transferred to a quantity of water whose temperature rise is measured.

Another often encountered form of specific heat is molar specific heat, c_m. It is defined as the heat added per *mole*[1] of material per degree Celsius. The relation between the quantity in moles of material undergoing a change of temperature, its temperature change, and associated heat transfer, which is equivalent to Eq. [13-1], is

[13-2] $\Delta Q = nc_m \Delta T$

In Eq. [13-2], n is the number of moles undergoing the change in temperature. It is found by dividing the mass of the substance by its molecular weight.

Ordinarily a pure substance exists in one of three states or *phases* of matter, liquid, solid, or vapor. There are exceptions, of course, such as amorphous or glassy solids. Some materials, like ice, appear in many different crystalline solid forms at high pressures. We will restrict our discussion to materials that exist in one of the three basic phases at each temperature and pressure experienced under experimental conditions. For such materials, a change of phase ordinarily takes place at *constant* temperature and pressure. Clearly, under these conditions, neither Eq. [13-1] nor [13-2] could correctly describe the transfer of heat accompanying the phase change, since both of these equations contain a factor ΔT. This factor is necessarily equal to zero for any process that occurs at constant temperature.

There is a definite quantity of heat evolved or absorbed in phase transitions of the kind being considered. The quantity of heat added to (or extracted from) 1 g of the substance undergoing a change of phase is referred to as the *latent* heat of the transition. For example, when 1 g of liquid water at 100°C and a pressure of 1 atm is converted to 1 g of steam at the same temperature and pressure, it is necessary that 539 cal be added to it. Conversely, when 1 g of steam at 100°C and 1 atm condenses to 1 g of liquid, 539 cal must be extracted from it. This quantity of heat is referred to as the latent heat of vaporization of water at 100°C and 1 atm (the normal boiling point of water). The heat that must be extracted from 1 g of liquid water at 0°C and 1 atm (its normal melting or freezing point) is called the latent heat of fusion of water. When m grams of a certain material undergo a phase transition, the associated heat transfer is given by

[13-3] $\Delta Q = mL$

[1] The usual definition of a mole, given in general chemistry, is Avogadro's number (6.02×10^{23}) of molecules, or a quantity of material whose mass in grams equals its molecular weight.

In Eq. [13-3], L represents the latent heat of the transition.

Experiments in thermal physics are generally performed with some sort of calorimeter. The simplest possible calorimeter consists of nothing more than a container into which to place the components of the system being studied, along with necessary auxiliary equipment such as thermometers, stirrers, etc. A very elementary experiment in calorimetry is illustrated in Fig. 13-1. Its purpose is to determine the specific heat c_s of a metal or other solid sample S. A thermometer T is used to read the initial temperature T_i of a mass m_w of water in the calorimeter. The calorimeter is labeled C in the diagram. The sample, whose mass is m_s, is placed in a steam bath for a sufficient time for it to come to a temperature of $100\,^\circ$C. It is removed from the steam bath and quickly dropped into the calorimeter. The final temperature of the system, T_f, is read after it has come to thermal equilibrium.

A simplified analysis of the experiment just described will serve to illustrate the type of calculations which are performed in calorimetry experiments. In addition to the data specified in the preceding paragraph, the mass and specific heat of the calorimeter (m_c and c_c) and of the thermometer (m_T and c_T) are assumed to be known. Any heat loss to (or gain from) the surroundings are going to be neglected in this analysis. In an actual experiment, careful attention must be paid to minimizing and/or correcting for heat exchange with the calorimeter environment.

The basic principle of physics upon which analysis of calorimetric data depends is conservation of energy. It takes the form of a *heat balance* (energy balance). The heat balance requires that the heat gain of those components of the system whose temperature rises or which absorb latent heat during an experiment is equated to the heat loss of the remaining components less any *net* heat loss to the surroundings). In this experiment, the only component that loses heat is the sample S. From Eq. [13-1], it

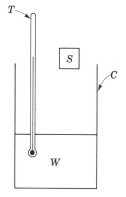

FIG. 13-1
Diagram of a simple calorimetric experiment to measure the specific heat of a metal sample (S). The apparatus consists of a calorimeter (C) containing water (W) in which a thermometer (T) has been immersed.

must lose $m_s c_s$ $(100\,^\circ\mathrm{C} - T_f)$ calories. The components gaining heat are the thermometer, the calorimeter, and the water inside it. The temperature rise of all these components is $T_f - T_i$. The heat balance equation may be written

[13-4] $(m_T c_T + m_c c_c + m_w c_w)(T_f - T_i) = m_s c_s (100 - T_f)$

The only unknown quantity in Eq. [13-4] is c_s, which may be determined by solving for it in the equation.

2. APPARATUS   In the foregoing discussion, we have conveniently neglected heat losses to (or gains from) the surroundings. Of course, in as simple an experiment as we have described, with a calorimeter completely exposed to and in thermal contact with the air around it, we would expect a substantial error in the calculated result due to heat exchange with the surroundings. In order to reduce the importance of this effect, calorimetric experiments are generally much more complicated in design. For instance, when measurements are performed at very low temperatures, heat exchanges with the surroundings must be kept to an absolute minimum. A complicated system of vacuum-walled calorimeters inside other vacuum-walled calorimeters, separated by layers of liquified gases, must then be used. The calorimeter that is to be used in this experiment is a good deal simpler than this.

Figure 13-2 shows a cross-section drawing of a double-walled calorimeter. The outer calorimeter cup C_1 is in thermal contact with the room. The inner calorimeter C_2 is separated by a volume of air from C_1. Since air is a relatively poor thermal conductor, C_2 is well isolated thermally from the outer calorimeter cup C_1

FIG. 13-2
Double-wall calorimeter apparatus, consisting of two calorimeters (C_1 and C_2), stirrer (S), and thermometer (T). Ice chips (I) may be added through a hole in the lid of the calorimeter apparatus.

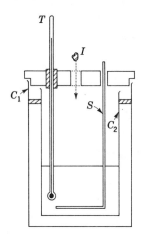

and the laboratory surroundings. Heat exchanges between the calorimeter and room are held to a fairly low value. Furthermore, this arrangement provides a well-defined *ambient* temperature, i.e., room temperature. This permits a quantitative heat loss correction to be performed. Details of this correction calculation will be discussed in Section 3.

The lid of the calorimeter is provided with three drilled holes. A rubber stopper, through which the thermometer T passes, fits in one of these. Another, smaller-diameter hole admits the handle of the metal stirrer S. A third hole is provided so that ice chips may be dropped periodically into the inner calorimeter. At other times this hole is covered by a lid or stopper. Ice chips must be wiped as dry as possible with an absorbent towel or cloth before they are added to the calorimeter. Any *liquid* water that remains on the surface of a chip will already have gained its latent heat of fusion from the room before it is introduced into the calorimeter. Its mass is included in the final weighing of the calorimeter. Therefore, it will have the effect of lowering the experimentally determined value of the latent heat of fusion.

A stopwatch is started the moment the first ice chip is added to the calorimeter. Measurements of the calorimeter temperature are made at regular intervals throughout the course of the experiment. Ice chips are added one by one, and the contents of the calorimeter are stirred continuously. In this way, a fairly regular time-temperature curve is obtained. When the temperature has been brought down to at least 10° below room temperature, the addition of ice is discontinued. The calorimeter temperature is measured at intervals for another 5 min in order to establish a warm-up line. The warm-up line in Figs. 13-3 and 13-4 is shown extrapolated back to the temperature axis. Its intercept on the T axis is labeled T_0 in both cases. The slope and intercept of the warm-up line are needed in the heat exchange analysis discussed in Section 3.

The latent heat of fusion of ice is calculated by means of a heat balance equation similar to Eq. [13-4]. In this case, there is no metal sample S that loses heat to the calorimeter and its contents. The latter lose heat instead to the ice that is added. In all other respects the calculation is the same.

3. THEORY   The physical basis for heat loss corrections is Newton's law of cooling. It states that the net rate of heat transfer (gain or loss) to the air surrounding a body is proportional to the difference

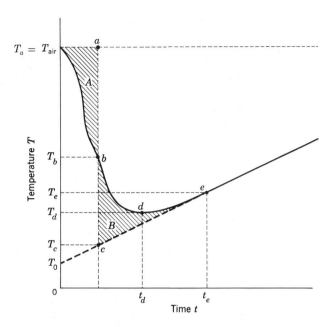

FIG. 13-3
Temperature-versus-time graph of an experiment to measure the heat of fusion of water. The determination of $T_a - T_c$ is discussed in the text. The line abc cuts off equal areas A and B.

in temperature between the body and the surrounding air. Expressed in equation form Newton's law of cooling is

$$[13\text{-}5] \quad \frac{dQ}{dt} = k(T - T_{\text{air}})$$

In this equation, Q represents heat *loss* from the body to the air around it, T is its instantaneous temperature, and k is a proportionality constant (which depends on the size, shape, and composition of the body). If the temperature of the body is lower than room temperature, the right-hand side of Eq. [13-5] is nega-

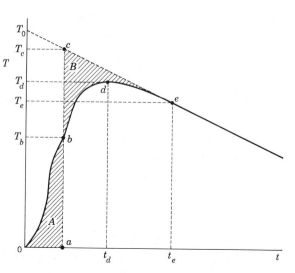

FIG. 13-4
The curve of Fig. 13-3 reflected about the room temperature line

tive, and, therefore, dQ/dt must also be negative. A negative dQ/dt is interpreted as a net rate of heat *gain* from the surroundings (i.e., a negative rate of heat loss to the surroundings). Thus, Eq. [13-5] holds for values of T greater than or less than T_{air}. A more convenient form of Eq. [13-5] for experiments in which the temperature is always below T_{air} is

[13-6] $$\frac{dQ}{dt} = k(T_{air} - T)$$

In Eq. [13-6], Q is understood to mean heat *gain* from the surroundings. It is the form of Newton's cooling law used here.

The integrated form of Newton's law of cooling is found by integrating both sides of Eq. [13-6] over a time interval from 0 to τ. When the integration has been performed, Eq. [13-6] becomes

[13-7] $$Q = k \int_{0}^{\tau} (T_{air} - T)\, dt$$

Equation [13-7] may be applied to a typical time-temperature (T-versus-t) graph such as Fig. 13-3. However, to give the equation more geometrical significance, it is convenient to reflect the graph about the horizontal line at $T = T_{air}$, i.e., to plot the $(T_{air} - T)$-versus-t graph instead. This is done in Fig. 13-4. According to Eq. [13-7], the area under this curve between 0 and τ must represent the total heat gain from the surroundings during this time interval (when it is multiplied by the proportionality constant k).

The significance of the heat exchange correction is that it enables us to determine an accurate value of $T_f - T_i$ to substitute in the heat balance equation required in this experiment. The difficulty encountered in determining the proper value of $T_f - T_i$ is apparent in Figs. 13-3 and 13-4. Should we take $T_a - T_e$ for this difference (referring to Fig. 13-3)? Should we use $(T_a - T_d)$ instead? Or should we take some other difference? The problem is that there is no sharp drop in temperature from T_i to T_f, but a rather gradual decrease in temperature as more ice is added to the calorimeter, followed by an even more gradual rise back toward room temperature after all the ice has been added.

The problem may be resolved by postulating an idealized experiment in which the temperature instantaneously drops from T_a to a final temperature, T_c in Figs. 13-3 and 13-4. The criterion which must be satisfied by the proper vertical straight line abc is that the total heat gain from the surroundings during the idealized experiment be equal to the heat gain during the actual experiment. In Figs. 13-3 and 13-4, this requirement is met by drawing abc at the point where the areas A and B are equal.[1]

In order to locate the proper time at which to position the vertical line abc in Fig. 13-4, the area under the actual time-temperature curve, i.e., the $(T_{air} - T)$-versus-t graph, must be known as a function of time. This implies that some numerical integration procedure be used to determine the value of the integral that appears in Eq. [13-7] as a function of time t. In the following section, several frequently used numerical integration techniques are described. These procedures are referred to as the trapezoidal rule and Simpson's rule. Their application to the heat-exchange correction will be discussed in Section 4.

Numerical Integration

Numerical integration is one of the most important applications of digital computers in scientific and engineering problems today. Only two elementary numerical integration methods will be discussed here, integration by means of the trapezoidal rule and by Simpson's rule. Of the two, although slightly more complicated, Simpson's rule is by far the more frequently employed numerical procedure. It is often preferred to more accurate (and more complicated) integration techniques because of its relative simplicity.

The trapezoidal rule is the most easily derived and the easiest to understand of all numerical integration procedures. It consists of approximating the actual area under a curve, $f(x)$, by the sum of the areas of small trapezoid-shaped elements into which the region has been divided. In Fig. 13-5 a curve, $y = f(x)$, is shown cut by vertical lines into elements of equal width, $x_{i+1} - x_i = h$. Trapezoids are formed when straight-line segments are drawn between adjacent points on the curve.

The height (ordinate) of the curve at x_i is $y_i = f(x_i)$. At $x_{i+1} = x_i + h$, the ordinate is $y_i = f(x_i + h)$. The formula for the area of the trapezoid between x_i and x_{i+1} is

[13-8] $A_i = \frac{1}{2}(x_{i+1} - x_i)(y_{i+1} + y_i) = \frac{1}{2}h(y_{i+1} + y_i)$

---

[1] The area A represents the heat gain during the actual experiment before the time at which the instantaneous temperature drop takes place in the idealized experiment. The area B represents the excess heat gain in the ideal experiment over and above the heat gain in the actual experiment, *after* the time at which the hypothetical instantaneous drop occurs. When these two areas are equal, the heat exchange with the surroundings is the same in the actual experiment as in its idealized counterpart.

FIG. 13-5
Diagram of a function
$y = f(x)$ divided into
$n - 1$ intervals for integration by the trapezoidal rule

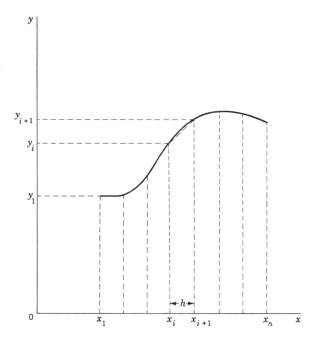

Adding up the areas of all the trapezoids, we find for the total area:

[13-9]

$$A = \sum_{i=1}^{n} A_i = \tfrac{1}{2}h(y_1 + y_2) + \tfrac{1}{2}h(y_2 + y_3) + \cdots + \tfrac{1}{2}h(y_{n-1} + y_n)$$

$$= \tfrac{1}{2}h(y_1 + 2y_2 + 2y_3 + \cdots + 2y_{n-1} + y_n)$$

Simpson's rule adds a nonlinear term to increase the accuracy of the trapezoidal approximation. It approximates the actual curve with a set of parabolas instead of straight-line segments. Simpson's rule may be applied to a function $f(x)$ which has been divided into $n - 1$ intervals of equal width h. It is performed by means of the formula

[13-10] $$A = \frac{h}{3}(y_1 + 4y_2 + 2y_3 + 4y_4 + \cdots + 2y_{n-2} + 4y_{n-1} + y_n)$$

The number of intervals must be *even*, i.e., n must be *odd*. Simpson's rule is discussed in greater detail in the numerical analysis references of the bibliography.

It is instructive to consider the results of the two methods and compare the accuracy of each when a simple function $f(x)$ is integrated using both procedures. Let $f(x)$ be the simple function x^4 for this comparison, and let the interval of integration be

$x = -2$ to $+2$. Furthermore, in order to provide a severe test of each procedure, let us choose a very coarse increment h in this case, $h = 1$. The curve $f(x)$ is thus divided into only four intervals. It is shown in Fig. 13-6.

Applying the trapezoidal rule, Eq. [13-9], to compute the area under the $f(x)$ curve, we obtain

$$A = \frac{1.0}{2}[16 + 2(1) + 2(0) + 2(1) + 16] = 18.00$$

Simpson's rule, Eq. [13-10], yields

$$A = \frac{1.0}{3}[16 + 4(1) + 2(0) + 4(1) + 16] = 13.33$$

The exact integration of this function leads to the result

$$A = \int_{-2}^{2} x^4\, dx = \left[\frac{x^5}{5}\right]_{-2}^{2} = \frac{1}{5}\Big[32 - (-32)\Big] = 12.80$$

Even with this coarse an increment, the area calculated by means of Simpson's rule is found to differ from the exact answer by only 4 percent. The trapezoidal method produces a numerical result which is about 40 percent greater than the actual value of the integral. In general, Simpson's rule requires many fewer intervals to attain a result with the same accuracy as is produced by the trapezoidal rule. Just how many fewer depends in part upon the particular type of function which is being integrated. The two procedures are compared, and the computational errors introduced by each are discussed in detail in most of the numerical analysis references listed in the bibliography.

FIG. 13-6
The area under the curve
$y = x^4$ **between** $+2$
and 2. The integral
$\int_{-2}^{2} x^4\, dx$
is determined by means
of the trapezoidal rule and
by Simpson's rule in the
text.

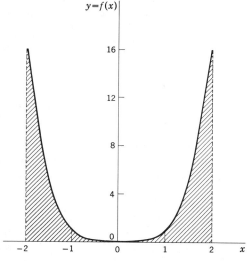

4. PROGRAMMING   The role of the computer here is to perform various numerical integrations which are required in the heat exchange correction for our calorimetry experiment. What the computer accomplishes, briefly, is the evaluation of certain graphical areas (integrals) in Fig. 13-4. These areas are related to the areas A and B. A knowledge of their values permits us to determine the correct time at which to draw the vertical line abc. This line cuts off equal A and B areas. As previously discussed, the height of the line abc between the time axis and the warm-up line is the correct value of ΔT to use in the heat balance equation.

The desired areas are shown in Fig. 13-8. It is the temperature-time graph of Fig. 13-4 broken into $n - 1$ intervals by equally spaced vertical lines. These vertical lines are h units (minutes or seconds) apart. They intersect the time axis at $t_1 = 0$, $t_2 = h$, $t_3 = 2h$, . . . , $t_i = (i - 1)h$, . . . , t_n; and the temperature-time curve, at the ordinates $T_1 = 0$, T_2, T_3, ...

Note that the vertical axis of Fig. 13-8 is labeled simply as T (it should be $T - T_{air}$), as it is in Fig. 13-4. When plotted, the raw experimental (T_i, t_i) data give rise to a graph like Fig. 13-3, not 13-4. In the initial step of the first DO loop of Fig. 13-7, these experimental T_i data are reflected about the $T = T_{air}$ line. They then take on the shape of the data plotted in Figs. 13-4 and 13-8. This reflection of T_i data takes place in the program step labeled

$$T_i = T_{air} - T_i$$

It corresponds to a Fortran statement like

T(I) = TAIR − T(I)

Verify that this statement produces the desired result.

The area under the temperature-time curve of Fig. 13-8 from $t = 0$ to $t = t_i$ is labeled A_{1_i} in Fig. 13-8. It is shown shaded with cross-hatched lines. The corresponding integral

[13-11] $A_{1_i} = \int_0^{t_i} T \, dt$

is computed numerically by means of the trapezoidal rule, as indicated in the flow chart of Fig. 13-7.

Values of the A_1 integral at successive points, t_1, t_2, . . . , t_i, . . . , t_n, are computed and stored in the elements of a one-dimensional array A1(I). Each value of the integral is computed from the previous one by means of a formula which is easily derived from the trapezoidal integration formula, Eq. [13-9]. Inspection of that equation shows that the difference between A_{1_i}

179

FIG. 13-7
Flow chart of a Fortran
program to compute the
heat exchange integrals
A_1, A_2, A_3, and A_4

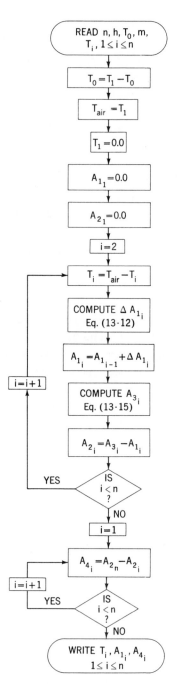

and $A_{1_{i-1}}$ is given by

[13·12] $\Delta A_{1_i} = A_{1_i} - A_{1_{i-1}} = \frac{1}{2}h(T_i + T_{i-1})$

This value is computed and added to the sum of all preceding ΔA_{1_i}'s to find A_{1_i}.

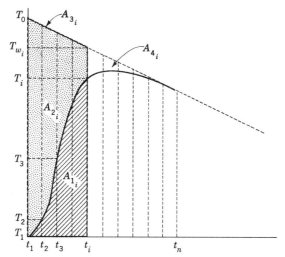

FIG. 13-8
Temperature-versus-time
curve of Fig. 13-4 broken
up into $n - 1$ intervals for
numerical integration. The
four integrals A_1, A_2, A_3,
and A_4 are evaluated
between $t = 0$ and
$t = t_i$.

The dotted area A_{2_i} in Fig. 13-8 is required to compute the area B in Figs. 13-3 and 13-4. The corresponding integral is most easily computed by subtracting A_{1_i} from the trapezoid-shaped area under the warm-up line in Fig. 13-8. This area, labeled A_{3_i}, may be computed with an expression derived from the area formula for a trapezoid. The derivation of this expression follows.

The warm-up line is a straight line with slope $-m$ and intercept T_0. Its equation is

[13-13] $T = T_0 - mt$

From Eq. [13-13], the ordinate of the warm-up line at t_i, i.e., T_{wi}, must be equal to $T_0 - mt_i$. Since $t_i = (i - 1)h$, T_{wi} must be given by

[13-14] $T_{wi} = T_0 - mh(i - 1)$

The ordinate of the warm-up line at t_1 is just T_0. From the formula for the area of a trapezoid, we have

[13-15] $A_{3_i} = \frac{1}{2}(t_i - t_1)(T_{wi} + T_0) = \frac{1}{2}(i - 1)h[2T_0 - mh(i - 1)]$

The integral A_{4_i} corresponds to the area between the temperature curve and warm-up line to the *right* of the vertical line at $t = t_i$. It is easily computed from A_{2_i} and A_{2_n}. A little consideration of Fig. 13-8 will confirm that A_{4_i} is just the difference between them, i.e.,

[13-16] $A_{4_i} = A_{2_n} - A_{2_i}$

Since A_{2_n} is not computed until the last pass through the first DO loop, Eq. [13-16] may not be used until that DO loop has been satisfied. Then, in a second computational DO loop, the A_4 integrals may be evaluated by means of this equation.

The A_1 integrals increase in magnitude with time, while the A_4's decrease with time. If both of these integrals are plotted together on the *same* time axis, they will intersect at some particular value of time. This is the correct value of time at which to draw the vertical line *abc* in Figs. 13-3 and 13-4. At this time the A_1 integral is identical to the shaded area A in Figs. 13-3 and 13-4; the A_4 integral, to B. (At the intersection point, A_1 is equal to A_4, which is the condition that A and B must satisfy). The height of the line *abc*, as previously discussed, is the correct value of ΔT to use in the heat balance equation for this experiment.

5. PROCEDURE   **a**   Weigh and record the mass of the dry calorimeter. Add to it about 100 g of water at room temperature. Weigh the calorimeter again and record its mass. Subtract the two readings to find the exact mass of water contained. Insert the thermometer and stirrer into the calorimeter.

b   Make sure the temperature of the calorimeter and its contents are the same as the temperature of the room before adding any ice. Break several ice cubes into small chips. Wrap the chips in an absorbent towel and let them come to 0°C. Dry each chip thoroughly with the towel before adding it to the calorimeter.

c   Start a clock timer at the instant the first chip is dropped into the inner calorimeter. Stir the calorimeter continuously and take temperature readings every 15 sec. Continue adding ice chips until the temperature has dropped at least 10° below room temperature. Keep recording the temperature at 30-sec intervals for 5 min after the last chip has been added. Weigh the calorimeter again. Determine the mass of ice added by subtracting the previously measured mass of the calorimeter (prior to the addition of ice).

d   Plot the temperature-versus-time curve for the calorimeter and its contents. Extrapolate the warm-up line back to the temperature axis. Determine both T_0 and the slope of the line. T_0 is its intercept on the T axis. These data are to be read into the computer by the program and analyzed by it.

e   Following the flow chart of Fig. 13-7, write a Fortran program to compute the integrals A_1, A_2, A_3, and A_4. With the aid of the computed integrals, locate the position of the vertical line that corresponds to line *abc* of Figs. 13-3 and 13-4. The ordinate of the warm-up line at the intersection with *abc* (i.e., T_c in Fig. 13-4) is the value of $T_f - T_i$ required in the heat balance equation. A good way to find this point is to plot both the A_1 and the A_4 integrals on the same time axis. Where the two curves intersect, the line abc is to be drawn. This procedure locates the position of the line accurately in between experimental points.

f Write a heat balance equation for this experiment similar to Eq. [13-4]. This equation should include two terms that account for the heat gain of the ice added:

$$m_i[L + c_w(T_f - 0)]$$

where m_i is the mass of ice and L is its latent heat of fusion. L is to be determined from the heat balance equation. The heat capacity of the thermometer is small; an approximate value for $m_T c_T$ of the thermometer will be supplied by your instructor. Corresponding terms for the calorimeter and stirrer may be evaluated with the aid of specific heat values taken from a handbook.

It is, of course, possible to write a computer program to perform the heat balance. The program requires the value of $T_f - T_i$ determined in step e. If so instructed, write an appropriate Fortran program for this analysis. A somewhat more elaborate single program may also be written which will (a) compute all integrals, (b) determine $T_f - T_i$ by interpolation, and (c) perform the heat balance analysis to find L.

1. THE ANALYTICAL BALANCE

The analytical balance is a delicate instrument used for the determination of the mass of an object.[1] The object is ordinarily placed on the left-hand pan of the balance, which is illustrated in Fig. A-1. The pan is labeled P in the figure. Precision weights in gram and milligram denominations are added to the right-hand pan until there is no tendency for the balance beam to tip in either direction when the pan supports are released. This is done by turning knob B. The pans are suspended from the upper beam (labeled A in Fig. A-1) which rests on a knife-edge support. If the left-hand pan drops when the pan supports are released, more

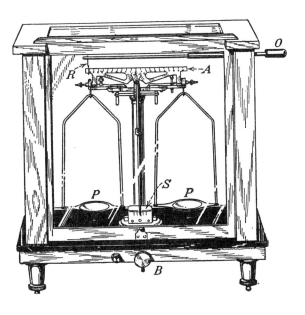

FIG. A-1
Analytical balance (*From "Experiments in Physics," 6th ed., by L. Ingersoll, M. Martin, and T. Rouse, McGraw-Hill Book Company, New York, 1953, with permission.*)

[1] The relation between the weight and mass of an object is $w = mg$. This equation is a consequence of Newton's second law, which is discussed in Experiment 6. Weight is measured in dynes and newtons (cgs and mks systems). Corresponding units of mass are grams and kilograms.

weight must be added to the right-hand pan, etc. Whenever the weights on the right-hand pan are changed, the pan supports should be brought to bear on the two pans to prevent damage to the delicate apparatus.

A fine balance is achieved by moving a light wire rider, R, along the upper beam until the pointer, once set in oscillation, swings through equal angles on either side of vertical. The balance condition is indicated on scale S by equal left and right excursions of the pointer. The rider may be moved by means of the externally manipulated rod labeled O in Fig. A-1. Most modern analytical balances are equipped with magnetic damping attachments that reduce the amplitude of oscillation of the balance beam to zero within a few vibrations.

The analytical balance is generally capable of weighing an object to a precision of 0.1 mg or better. The more rugged, double-pan balances, or trip scales, described below often may be read to the nearest 0.01 g, although sometimes only to 0.1 g. The principle of operation of the two kinds of balance is exactly the same. Both types of balance depend on the principle of the lever.

The balances shown in Figs. A-1 and A-2 are nothing more than levers with equal arm lengths. When equal forces are applied to the ends of both lever arms of either balance, the system is put in mechanical equilibrium. In equilibrium, there is no tendency for it to move, if undisturbed. The equal forces on the two arms of a balance are the gravitational (weight) forces acting on the equal masses which have been placed on its left- and right-hand pans.

FIG. A-2
Trip scale (balance)

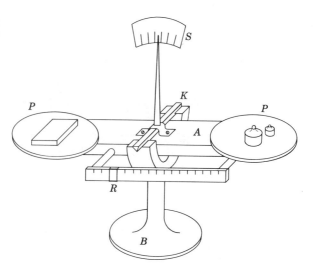

2. PAN BALANCE (TRIP SCALE)

A simple double-pan balance, or trip scale, is illustrated in Fig. A-2. It consists of two pans, both labeled P in the figure, which rest on the horizontal beam A. The beam is balanced on a knife edge, which is labeled K in the figure. Attached to the beam is a pointer which may be viewed against scale S to indicate the orientation of the beam. As in the case of the analytical balance, the object to be weighed is placed on the left-hand pan; weights are added to the right-hand pan until a coarse balance is achieved. The rider R is moved to the right until a fine balance is obtained. The principle of operation of the pan balance is the same as that of the analytical balance.

3. THE VERNIER CALIPER

The construction of a simplified vernier caliper is illustrated in Fig. A-3a. The upper scale, or *main scale,* is ruled into millimeter divisions. The length of the scale shown in the figure is 2.0 cm. The scale on the lower, movable jaw of the caliper is called a *vernier.* The purpose of the vernier is to divide each upper scale division *precisely* into ten smaller divisions, i.e., to permit measurements to be made accurately to the nearest tenth of a millimeter. As may be seen in the figure, 10 vernier divisions coincide exactly with nine millimeter divisions of the main scale. Thus, each vernier division is 0.9 mm wide. The zero marks on the two scales coincide when the jaws of the caliper are closed.

FIG. A-3
Principle of the vernier caliper. Ten vernier divisions are equal to nine main scale divisions (a); a 12.2-mm cylinder being measured (b).

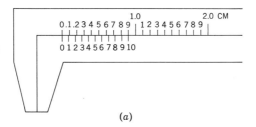

(a)

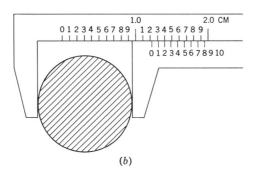

(b)

The operation of the vernier is illustrated in Fig. A-3b. It is shown measuring a cylindrical object 12.2 mm (1.22 cm) in diameter. The movable jaw must be opened by that amount to permit the cylinder to fit snugly between the jaws. When the jaws have been opened to this extent, the zero on the lower scale lands somewhere between the 12- and 13-mm divisions on the main scale. The whole number of millimeters in the diameter of the cylinder is thus equal to 12. The number of tenths of a millimeter is indicated by the vernier scale. It is equal to the number associated with that division on the vernier scale which comes *closest* to lining up with *some* number on the main scale. In the case illustrated in Fig. A-3b the correct number of tenths to add is 2 since the vernier numeral 2 lines up with a division mark (the 14 mm division) on the main scale.

That the diameter of the cylinder is 12.2 mm is easily proved. The division numbered 2 on the vernier scale coincides with the 14-mm division on the upper scale. Thus, the movable lower jaw must have been opened by 14.0 mm less the width of the two vernier divisions (located between 0 and 2 on the vernier). Saying it another way, 14.0 mm on the main scale must correspond to the width of two divisions of the vernier scale plus the diameter of the cylinder. This may be seen clearly in Fig. A-3b. The two vernier divisions occupy 2×0.9 mm $= 1.8$ mm. By difference, therefore, the diameter of the cylinder must be 14.0 mm $-$ 1.8 mm $= 12.2$ mm.

The reading of a vernier caliper may be summarized as follows:

1   Find the integral number of main-scale divisions (millimeters, in this case) in the length of the measured object by counting the number of main-scale divisions to the left of zero on the vernier scale.

2   Add to this length the number of tenths indicated by that number on the vernier scale which lines up with *any* number on the main scale. If no vernier division lines up exactly with a main-scale division, find the vernier division that most nearly does align itself with a main-scale division. Add this number of tenths to the length found in step 1.

A vernier may be constructed to divide one division on the main scale into any number of equal parts. Let N be the desired number. The vernier is then ruled so that N of its divisions correspond to *exactly* $N - 1$ divisions on the main scale.

4. MICROMETER CALIPER

The construction of a metric micrometer is illustrated in Fig. A-4. It is very similar in operation to the vernier caliper. In place of

188

FIG. A-4
Micrometer caliper (*From*
"Experiments in Physics,"
6th ed., by L. Ingersoll,
M. Martin, and T. Rouse,
McGraw-Hill Book Company,
New York, 1953, with
permission.)

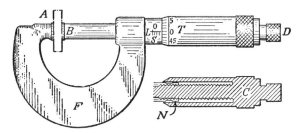

a linear vernier scale, it has one inscribed on its rotating barrel. The barrel is permanently fastened to a precision screw thread, as shown in the inset drawing. The screw travels precisely 0.5 mm in one complete turn, i.e., its *pitch* is 0.500 mm. The vernier scale on the barrel is ruled into 50 divisions. Therefore, each division corresponds to an increment of 0.01 mm in the length of the measured object.

The micrometer screw passes through the frame F of the micrometer, and terminates in a machined jaw B. The frame is attached to a fixed jaw A. The object to be measured is placed between the jaws, and the screw is tightened down on it. In order that the micrometer jaws always exert the same pressure on measured objects, the micrometer screw is provided with a ratchet device D, which begins to slip once the desired pressure has been attained. This device also protects the micrometer from being damaged by excessive tightening.

Corresponding to the main scale on the vernier caliper is the scale S on the micrometer. It is affixed to the frame. The micrometer screw passes through a threaded channel under scale S. It is ruled with 0.5-mm divisions. The whole number of half-millimeters in the length of a measured object is indicated by the number of divisions on scale S which lie to the left of the barrel scale T. Three such divisions are seen in Fig. A-4.

The number of *hundredths* of a millimeter that must be added to the length of the measured object is indicated by the division on scale T which last turned past scale S, e.g., 49 in Fig. A-4. Including the estimated tenths of a division of scale T, the illustrated micrometer reading is 1.999 mm. This reading is made up of three half-millimeters, forty-nine 0.01-mm divisions on T, and an estimated additional nine-tenths of a T division. In equation form, $l = 3(0.5 \text{ mm}) + 49(0.01 \text{ mm}) + 0.9(0.01 \text{ mm}) = 1.999$ mm.

APPENDIX B
THE AIR TRACK—
MEASUREMENT OF g

The linear air track is a useful device for studying one-dimensional motion in the absence of friction.[1] In fact, friction cannot be totally eliminated from any mechanics experiment performed in the laboratory. However, by suspending a moving car (or glider) on a thin film of air, the frictional force between car and track is greatly reduced, and, to a high degree of approximation, may be neglected in most calculations.

The air track apparatus is illustrated in Fig. B-1. It consists of a square or triangular, hollow extruded aluminum tube. The tube is mounted on a horizontal support frame with a 90° corner pointing, upward, i.e., each of the sides of the air track make an angle of 45° with vertical. In this way a track with the shape of an inverted V is formed. Several rows of small holes are drilled in the walls of the V-shaped track. One end of the hollow tube is sealed; the other is provided with a coupling and is connected to an air

FIG. B-1
Air-track apparatus (*After the Stull-Ealing linear air track, with permission of The Ealing Corporation.*)

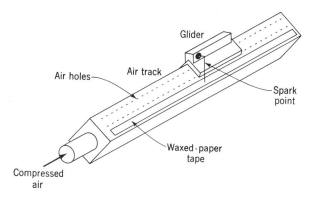

Glider

Air holes

Air track

Spark point

Waxed-paper tape

Compressed air

[1] The air-track apparatus described here and pictured in Fig. B-1 is similar to the linear air trough introduced by H. V. Neher and R. B. Leighton (*Am. J. Phys.*, vol. 31, p. 255, April, 1963) and modified by J. L. Stull (*Am. J. Phys.*, vol. 30, p. 839, November, 1962). The commercial version of this equipment is available from the Ealing Corporation, Cambridge, Mass. The Ealing air track has a triangular rather than square cross section.

compressor. Compressed air at a small positive pressure is applied to the tube. Air flowing through the several rows of holes provides the film on which the glider rides with negligible friction.

The glider is constructed of a short section of aluminum angle which fits over the track. A platform to which weights may be attached is fastened to the top of the angle. The glider may be provided with a wire spark point which is connected to a high-voltage synchronous spark timer. The spark point is adjusted so that it travels directly above a waxed-paper tape resting on the side of the air track. As the spark timer discharges periodically, it causes a row of spark impressions to be made on the waxed tape.

An experiment to determine the gravitational acceleration constant with the air track apparatus is illustrated in Fig. B-2. The track is first leveled, and then inclined at a small angle θ with horizontal. The elevation angle θ may be calculated from measurements of the length l of the air track and the height h to which one side of the track is raised. It may be calculated with the aid of the trigonometric relation defining the sine of θ

[B-1] $\quad \sin \theta = \dfrac{h}{l}$

The forces **N, w,** and **f** acting on the glider are illustrated in Fig. B-2. The frictional force **f,** as previously discussed, is very small and may be neglected in the calculation of g. The normal force, labeled **N,** acts perpendicular to the track. The weight force **w** is shown resolved into components parallel to and perpendicular to the track. These components are labeled w_x and w_y (the x axis is taken parallel to and pointing down the track; the y axis, perpendicular to the track and pointing upward). The magnitudes of these components are

[B-2] $\quad w_x = mg \sin \theta$

and

[B-3] $\quad w_y = mg \cos \theta$

FIG. B-2
Experimental arrangement of the air track to determine the gravitational acceleration constant
(*After the Stull-Ealing linear air track, with permission of The Ealing Corporation.*)

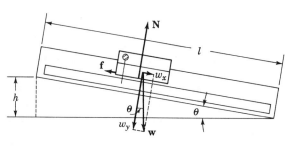

These equations follow from Eqs. [3-3] and [3-4] and from the relation between weight and mass, $w = mg$.

The measured acceleration of the glider may be related to the gravitational acceleration constant g by means of Newton's second law. Neglecting the small friction force f we find only one component of force which acts parallel to the incline. Applying the second law in its x-component form, we have

[B-4]　　$F_x = mg \sin \theta = ma_x$

Dividing Eq. [B-4] by m, and solving for g, we obtain

[B-5]　　$g = \dfrac{a_x}{\sin \theta}$

The experimental a_x is obtained from the spark-tape data in the same manner as the gravitational acceleration is computed in Experiment 4 from the spark records made with the free-fall apparatus. The sine of the elevation angle θ may be determined from Eq. [B-1]. Equation [B-5] may be used to calculate g once a_x and θ have been computed.

A TRIGONOMETRIC METHOD
OF ADDITION OF VECTORS

We consider in this section a trigonometric method for the addition of three vectors. The method is based upon the polygon method of addition of three or more vectors. It is described in detail in most of the general physics references in the bibliography. The polygon construction is illustrated in Fig. C-1. The three vectors which are to be added are identical to the vectors shown in Fig. 3-4, where these vectors are labeled **A**, **B**, and **C**. They make angles of zero, β, and δ with the x axis (the x axis is chosen to be coincident with vector **A**).

In Fig. C-1 the resultant of **A** and **B** is assigned the magnitude APB (standing for A plus B); it makes an angle ϕ with the x axis. The resultant of all three vectors is denoted by **APBPC;** it makes an angle θ with the x axis. Applying the cosine law to the lower triangle, which contains **A, B** and **APB,** we have for the magnitude of the resultant of **A** and **B**

[C-1] $APB = (A^2 + B^2 - 2AB \cos \beta')^{1/2}$

Since β and β' are supplementary, and therefore have cosines which differ only by a minus sign, we obtain finally for APB

[C-2] $APB = (A^2 + B^2 + 2AB \cos \beta)^{1/2}$

FIG. C-1
The polygon method of addition of three vectors A, B, and C

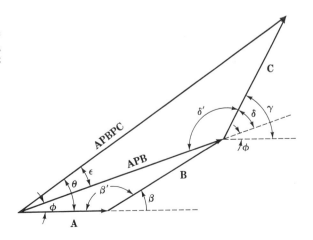

To compute the angle ϕ, we apply the law of sines to the same triangle

[C-3] $$\frac{\sin \phi}{B} = \frac{\sin (\pi - \beta)}{APB}$$

Solving for ϕ, we obtain

[C-4] $$\phi = \sin^{-1}\left[\frac{B \sin (\pi - \beta)}{APB}\right]$$

When we come to solve for the magnitude of the resultant of all three vectors, we first must compute the angle δ from the given angle γ and the previously computed angle ϕ

[C-5] $$\delta = \gamma - \phi$$

This relation follows from an inspection of Fig. C-1, where δ has been drawn as the angle between **C** and the extension of **APB**. This angle bears the same relation to the upper triangle in the figure that β does to the lower triangle. Therefore, by analogy to Eq. [C-2] we may write for the magnitude of the resultant

[C-6] $$APBPC = (C^2 + APB^2 + 2C\ APB\ \cos \delta)^{1/2}$$

The angle ϵ is the analog of ϕ. By analogy to Eq. [4-8] we may write

[C-7] $$\epsilon = \sin^{-1}\left[\frac{C \sin (\pi - \delta)}{APBPC}\right]$$

Finally, referring to Fig. C-1 once again, we see that the desired angle θ is equal to the sum of ϵ and ϕ

[C-8] $$\theta = \epsilon + \phi$$

A Fortran program based on Eqs. [C-1] through [C-8] may readily be written. Apart from input and output statements, it should consist of a sequence of Fortran arithmetic statements which are equivalent to these equations. In fact, a program almost identical to the one presented in Fig. D-1 may be used here. Only its arithmetic statements need to be modified in the manner described in this section. It is an instructive exercise to write such a program.

SAMPLE FORTRAN PROGRAMS

The following Fortran programs are associated with the flow charts presented in Experiments 3 through 7. Each program performs an analysis of the data taken in the corresponding experiment.

FIG. D-1
Program for Experiment 3 to add vectors A, B, and C

```
C ADDITION OF THREE VECTORS
      READ (5,100) A, B, C, BETA, GAMMA
      BETA = BETA/57.29578
      GAMMA = GAMMA/57.29578
      ALPHA = 0.0
      AX = A
      AY = 0.0
      BX = B*COS(BETA)
      BY = B*SIN(BETA)
      CX = C*COS(GAMMA)
      CY = C*SIN(GAMMA)
      RX = AX + BX
      RY = AY + BY
      APB = SQRT(RX**2 + RY**2)
      PHI = 57.29578*ATAN(RY/RX)
      RX = AX + BX + CX
      RY = AY + BY + CY
      APBPC = SQRT(RX**2 + RY**2)
      THETA = 57.29578*ATAN(RY/RX)
      BETA = BETA*57.29578
      GAMMA = GAMMA*57.29578
      WRITE (6,100) A, B, C, APB, APBPC, ALPHA, BETA, GAMMA, PHI, THETA
100 FORMAT (5F15.4)
      END
```

FIG. D-2
Program for Experiment 4 to analyze the free-fall spark-tape data

```
C     CALCULATION OF G
C     METHOD OF DIFFERENCES
C
      REAL T(20), Y(20), V(20), A(20)
      READ (5,100) N, DELT
      I = 1
    4 READ (5,101) Y(I)
      I = I + 1
      IF (I - N) 4, 4, 7
    7 T(1) = 0.0
      V(1) = (Y(2) - Y(1))/DELT
      I = 2
    8 T(I) = T(I-1) + DELT
      V(I) = (Y(I+1) - Y(I))/DELT
      A(I) = (V(I) - V(I-1))/DELT
      I = I + 1
      IF (I - N) 8, 12, 12
   12 T(N) = T(N-1) + DELT
      WRITE (6,102) T(1), Y(1)
      WRITE (6,103) V(1)
      I = 2
   16 WRITE (6,102) T(I), Y(I), A(I)
      WRITE (6,103) V(I)
      I = I + 1
      IF (I - N) 16, 20, 20
   20 T(N) = T(N-1) + DELT
      WRITE (6,102) T(N), Y(N)
100 FORMAT (I2,F10.3)
101 FORMAT (F10.3)
102 FORMAT (2F10.3,F20.3)
103 FORMAT (F30.3)
      END
```

FIG. D-3
Program for Experiment 5
to compute centripetal
force by means of the
method of differences

```
C       CENTRIPETAL FORCE CALCULATION
C       METHOD OF DIFFERENCES
C
        REAL REV(50), M
      2 READ (5,99) N, DELT, RAD, M
        I = 1
      3 READ (5,100) REV(I)
        I = I + 1
        IF (I - N) 3, 3, 5
      5 WRITE (6,101) DELT, RAD, M
        IMAX = N/2
        SUM = 0.0
        I = 1
      7 DREV = REV(IMAX + I) - REV(I)
        SUM = SUM + DREV
        I = I + 1
        IF (I - IMAX) 7, 7, 12
     12 RPS = 4.0*SUM/(DELT*FLOAT(N**2))
        FC = 4.0*M*RAD*(3.14159*RPS)**2
        I = 1
     16 WRITE (6,102) I, REV(I)
        I = I + 1
        IF (I - N) 16, 16, 19
     19 WRITE (6,101) RPS, FC
        GO TO 2
     99 FORMAT (I2,3F10.4)
    100 FORMAT (F10.4)
    101 FORMAT (3F14.4)
    102 FORMAT (I4,F12.0)
        END
```

FIG. D-4
Program for Experiment 6
to analyze spark-tape data

```
        DIMENSION X(10,50), V(50), A(50), T(50), W(10), ASUM(10)
        READ (5,100) N, M, DELT, G, WO
        DO 4 I = 1, N
      4 READ (5,101) (X(I,J), J = 1,M), W(I)
        MM1 = M - 1
        T(1) = 0.0
        DO 14 I = 1, N
        ASUM(I) = 0.0
        WRITE (6,102) W(I)
        V(1) = (X(I,2) - X(I,1))/DELT
        WRITE (6,102) T(1), X(I,1)
        DO 12 J = 2, MM1
        T(J) = T(J-1) + DELT
        V(J) = (X(I,J+1) - X(I,J))/DELT
        A(J) = (V(J) - V(J-1))/DELT
        ASUM(I) = ASUM(I) + A(J)
     12 WRITE (6,102) T(J), X(I,J), V(J-1), A(J)
        T(M) = T(MM1) + DELT
     14 WRITE (6,102) T(M), X(I,M), V(MM1)
        XM = M - 2
        DO 15 I = 1,N
        AEX = ASUM(I)/XM
        ATH = W(I)*G/(W(I) + WO)
     15 WRITE (6,102) W(I), AEX, ATH
    100 FORMAT (2I3,3F10.0)
    101 FORMAT (8F10.0)
    102 FORMAT (4F14.4)
        END
```

FIG. D-5
Program for Experiment 7
to analyze rotational
motion

```
C     ROTATIONAL MOTION PROBLEM
C
C     READ IN/PRINT OUT DATA AND PARAMETERS
      REAL M1, M2, IO, M(20), T(20), ALPHA1(20), ALPHA2(20), IOR      001
      READ (5,100) N, M1, M2, R1, R2, F, G, H                         002
      WRITE (6,101) M1, M2, R1, R2, F, G, H                           003
      WRITE (6,102)                                                   004
      I = 1                                                           005
   10 READ (5,103) M(I), T(I)                                         010
      WRITE (6,104) I, M(I), T(I)                                     011
      I = I + 1                                                       012
      IF (I - N) 10, 10, 15                                           013
   15 WRITE (6,105)                                                   015
C     CALCULATE ANGULAR ACCELERATIONS - PRINT OUT ALPHAS
      I = 1                                                           016
      C = 2.0*H/R1                                                    017
      IO = 0.5*(M1*R1**2 + M2*R2**2)                                  018
      IOR = IO/R1                                                     019
   20 ALPHA1(I) = C/T(I)**2                                           020
      ALPHA2(I) = (M(I)*G - F)/(IOR - R1*M(I))                        021
      WRITE (6,104) I, M(I), ALPHA1(I), ALPHA2(I)                     022
      I = I + 1                                                       023
      IF (I - N) 20, 20, 25                                           024
   25 WRITE (6,106) IO
C     FORMAT STATEMENTS
  100 FORMAT (I2/(2F10.0))                                            100
  101 FORMAT (1H1,4H M1=, F9.5,4H M2=, F9.5,4H R1=, F7.5,4H R2=, F7.5, 101
     1 3H F=, F7.5,3H G=, F7.5,3H H=, F7.5)                          101.1
  102 FORMAT (32H0TRIAL LOADMASS(KG)   TIME(SEC) )                     102
  103 FORMAT (2F10.0)                                                 103
  104 FORMAT (1H ,I4,4F13.4)                                          104
  105 FORMAT (46H0TRIAL LOADMASS(KG)   ALPHA,EXPER. ALPHA,THEOR.)      105
  106 FORMAT (27H0 MOMENT OF INERTIA (KG-M2)/15X,E12.5)
      END
```

BIBLIOGRAPHY

A. GENERAL PHYSICS REFERENCES

Alonso, M., and **E. J. Finn:** "Physics," Addison-Wesley Publishing Company, Inc., Reading, Mass., 1970.

Borowitz, S., and **A. Beiser:** "Essentials of Physics," Addison-Wesley Publishing Company, Inc., Reading, Mass., 1966.

Bueche, F.: "Introduction to Physics for Scientists and Engineers," McGraw-Hill Book Company, New York, 1969.

Feynman, R. P., R. B. Leighton, and **M. Sands:** "The Feynman Lectures on Physics," vol. 1, Addison-Wesley Publishing Company, Inc., Reading, Mass., 1963.

Fowler, R. G., and **D. I. Meyer:** "Physics for Scientists and Engineers," 2nd ed., Allyn and Bacon, Inc., Boston, 1962.

Freier, G. D.: "University Physics: Experiment and Theory," Appleton-Century-Crofts, Inc., New York, 1965.

Halliday, D., and **R. Resnick:** "Fundamentals of Physics," John Wiley & Sons, Inc., New York, 1970.

Haxen, W. E., and **R. W. Pidd:** "Physics," Addison-Wesley Publishing Company, Reading, Mass., 1965.

Kingsbury, R. F.: "Elements of Physics," D. Van Nostrand Company, Inc., Princeton, N.J., 1965.

Kittel, C., W. D. Knight, and **M. A. Ruderman:** "Berkeley Physics Course—Volume I—Mechanics," McGraw-Hill Book Company, New York, 1965.

Michels, W. C., M. Correll, and **A. L. Patterson:** "Foundations of Physics," D. Van Nostrand Company, Inc., Princeton, N.J., 1968.

Morgan, J.: "Introduction to University Physics," 2nd ed., Allyn and Bacon, Inc., Boston, 1969.

Sears, F. W., and **M. W. Zemansky:** "University Physics," 4th ed., Addison-Wesley Publishing Company, Inc., Reading, Mass., 1970.

Semat, H.: "Fundamentals of Physics," 4th ed., Holt, Rinehart and Winston, Inc., New York, 1966.

Shortley, G. H., and **D. E. Williams:** "Elements of Physics," 4th ed., Prentice-Hall, Inc., Englewood Cliffs, N.J., 1965.

Weber, R. L., M. W. White, and **K. V. Manning:** "Physics for Science and Engineering," McGraw-Hill Book Company, New York, 1959.

Weidner, R. T., and **R. L. Sells:** "Elementary Classical Physics," Allyn and Bacon, Inc., Boston, 1965.

Young, H. D.: "Fundamentals of Mechanics and Heat," McGraw-Hill Book Company, New York, 1964.

B. REFERENCES ON EXPERIMENT DESIGN AND ERROR ANALYSIS

Baird, D. C.: "Experimentation: An Introduction to Theory and Experiment Design," Prentice-Hall, Inc., Englewood Cliffs, N.J., 1962.

Barford, N. C.: "Experimental Measurements: Precision, Errors and Truth," Addison-Wesley Publishing Company, Inc., Reading, Mass., 1967.

Beers, Y.: "Introduction to the Theory of Error," 2nd ed., Addison-Wesley Publishing Company, Inc., Reading, Mass., 1957.

Braddick, H. J.: "The Physics of the Experimental Method," 2nd ed., Chapman and Hall, Ltd., London, 1966.

Brinkworth, B. J.: "An Introduction to Experimentation," American Elsevier Publishing Company, Inc., New York, 1968.

Fry, T. C.: "Probability and Its Engineering Uses," 2nd ed., D. Van Nostrand Company, Inc., Princeton, N.J., 1966.

Parratt, L. G.: "Probability and Experimental Errors in Science," John Wiley & Sons, Inc., New York, 1961.

Topping, J.: "Errors of Observation and Their Treatment," 3rd ed., The Institute of Physics, London, 1966.

Wilson, E. B.: "An Introduction to Scientific Research," McGraw-Hill Book Company, New York, 1952.

Young, H. D.: "Statistical Treatment of Experimental Data," McGraw-Hill Book Company, New York, 1962.

C. PROGRAMMING AND NUMERICAL ANALYSIS REFERENCES

Arden, B. W.: "An Introduction to Digital Computing," Addison-Wesley Publishing Company, Inc., Reading, Mass., 1963.

Bork, A. M.: "FORTRAN for Physics," Addison-Wesley Publishing Company, Inc., Reading, Mass., 1967.

Dimitry, D. L., and **T. H. Mott:** "Introduction to FORTRAN IV Programming," Holt, Rinehart and Winston, Inc., New York, 1966.

Farina, M. V.: "FORTRAN IV Self-Taught," Prentice-Hall, Inc., Englewood Cliffs, N.J., 1966.

Golde, H.: "FORTRAN II and IV for Engineers and Scientists," The Macmillan Company, New York, 1966.

Golden, J. T.: "FORTRAN IV Programming and Computing," Prentice-Hall, Inc., Englewood Cliffs, N.J., 1965.

Haag, J. N.: "Comprehensive FORTRAN Programming," Hayden Book Company, Inc., New York, 1965.

Harvill, J. B.: "Basic FORTRAN Programming," 2nd ed., Prentice-Hall, Inc., Englewood Cliffs, N.J., 1968.

Harris, L. D.: "Numerical Methods Using FORTRAN," Charles E. Merrill Company, New York, 1964.

Hull, T. E.: "Introduction to Computing," Prentice-Hall, Inc., Englewood Cliffs, N.J., 1966.

Jacquez, J. A.: "A First Course in Computing and Numerical Methods," Addison-Wesley Publishing Company, Inc., Reading, Mass., 1970.

Jamison, R. V.: "FORTRAN Programming," McGraw-Hill Book Company, New York, 1966.

Kuo, S. S.: "Numerical Methods and Computers," Addison-Wesley Publishing Company, Inc., Reading, Mass., 1965.

Lecht, C. P.: "The Programmer's FORTRAN II and FORTRAN IV," McGraw-Hill Book Company, New York, 1966.

Ledley, R. S.: "FORTRAN IV Programming," McGraw-Hill Book Company, New York, 1966.

Lee, R. M.: "A Short Course in FORTRAN IV Programming," McGraw-Hill Book Company, New York, 1967.

McCalla, T. R.: "Introduction to Numerical Methods and FORTRAN Programming," John Wiley & Sons, Inc., New York, 1967.

McCormick, J. M., and **M. G. Salvadori:** "Numerical Methods in FORTRAN," Prentice-Hall, Inc., Englewood Cliffs, N.J. 1964.

McCracken, D. D.: "A Guide to FORTRAN IV Programming," John Wiley & Sons, Inc., New York, 1965.

McCracken, D. D., and **W. S. Dorn:** "Numerical Methods and FORTRAN Programming," John Wiley & Sons, Inc., New York, 1964.

Murrill, P. W., and **C. Smith:** "FORTRAN IV Programming for Engineers and Scientists," International Textbook Company, Scranton, Pa., 1968.

Nydegger, A. C.: "An Introduction to Computer Programming with an Emphasis on FORTRAN IV," Addison-Wesley Publishing Company, Inc., Reading, Mass., 1968.

Organick, E. I.: "A FORTRAN IV Primer," Addison-Wesley Publishing Company, Inc., Reading, Mass., 1966.

Pennington, R. H.: "Introductory Computer Methods and Numerical Analysis," The Macmillan Company, New York, 1965.

Pollack, S. W.: "A Guide to FORTRAN IV Programming," Columbia University Press, New York, 1965.

Prager, W.: "Introduction to Basic FORTRAN Programming and Numerical Methods," Blaisdell Publishing Company, New York, 1965.

Ralston, A.: "A First Course in Numerical Analysis," McGraw-Hill Book Company, New York, 1965.

Southworth, R. W., and **S. L. Deleeuw:** "Digital Computation and Numerical Methods," McGraw-Hill Book Company, New York, 1965.

Wilf, H. S.: "Programming for a Digital Computer in the FORTRAN Language," Addison-Wesley Publishing Company, Inc., Reading, Mass., 1969.

INDEX